ROCKSTAR

From the Shadows of the Crack Epidemic to the Light of Redemption

by George Cortez Johnson

TABLE OF CONTENTS

Preface

There's a rawness to the streets where I grew up—a kind of electric tension that hums under the cracked sidewalks, thick enough to taste on your tongue if you're close enough. It's the kind of place where each breath can be a gift and every decision a matter of life or death. The air itself seems charged with fractured dreams, whispered betrayals, and desperate hope, all tangled in the smoky haze of nights that stretch too long and days that never quite bring relief. This is not a tale spun from distant fantasy or abstract sorrow; it is a visceral account born from the very core of that gritty, unforgiving world—a world shaped by the shadow of the crack epidemic, raw gang violence, and the suffocating weight of addiction. But beyond the darkness lurks something unexpected: a story of survival, of stubborn resilience, and eventually, of redemption.

ROCKSTAR is the name I was given, or maybe the title I earned by falling hard and rising slower still. It's a nickname that feels heavier than just an epithet, a paradox that reflects both the chaos and the flicker of light threading through my journey. This book isn't just about the wreckage wrought by the crack epidemic—a plague that infiltrated everything from playgrounds to church pews, from whispered deals on street corners to the brutal calculus of survival inside cramped jail cells. It's about a young man caught in the maelstrom of that era, stumbling into gang life like it was the only escape from hunger and fear, then spiraling into addiction that nearly swallowed him whole. But it's also about the moments that pulled him back, moments so fragile and fierce they became the bedrock for the fight to reclaim a fractured life.

Writing this book meant laying bare the raw, unvarnished realities I lived through—because silence and sugarcoating have never helped anyone. The crack epidemic didn't just decay neighborhoods; it fractured families, crushed innocence, and bred a cycle of violence so relentless it seemed unstoppable. But there is an urgent need to understand what happens behind the headlines and statistics—to feel the trembling hands of a kid desperate to outrun poverty, the thunderous roar of bullets in alleyways, the agonizing grip of addiction tightening around a person's soul. This life, my life, was carved out in the storm of those forces. Yet within that storm, the human spirit pulses with a fierce resistance, refusing to be extinguished even when all odds say it should be.

So, this memoir is more than just my story—it's a mirror held up to a society wrestling with pain cloaked in stigma, a call to look beyond stereotypes and judgments. It's about the cost of systemic neglect and the power of empathy, the threads that bind pain to healing, chaos to hope. The military service I sought after incarceration was a paradox of discipline and trauma; the fragile love I found with a woman exposed me to vulnerability, yet threatening to unravel everything I'd barely tethered together. Homelessness was not just physical displacement but a crucible where identity and despair collided. And recovery? That tangled, grueling path through therapy rooms, group meetings, and endless self-confrontation wasn't a neat redemption arc but a continuous wrestle with darkness and light.

Each chapter you will read is crafted to carry you through these layers— sometimes a pounding rush of violence and desperation, then softening into quiet reflections where voices of family, love, and advocacy echo. I wanted these pages to feel alive, uneven like life itself, with rhythms that pull you down into the grimiest trenches and lift you toward moments of revelation and grace. The narrative voice you'll encounter is candid and unrelenting, sometimes raw to the bone because that is what truth demands, but it also breathes with a lyrical pulse that honors those fleeting sparks of beauty found even in the bleakest nights.

I invite you to step into this world with open eyes and an open heart. This book does not offer easy answers or sanitized pain; instead, it offers a plunge into a complex reality—where addiction meets survival, where love is both a wound and a salve, where advocacy emerges from the ashes of suffering. It is a testament to the resilience that can emerge from the margins, to the courage that whispers in the dark, fighting to turn brokenness into transformation.

If you come seeking redemption as a clean, polished destination, be prepared instead for a journey marked by stumbles and setbacks, but also small victories that mean everything. If you come prepared to meet the harsh honesty of struggle, you'll find, I hope, that beneath the grime and fury lies the enduring human will to heal and to rise. ROCKSTAR is my story, but it is also a rallying cry for all who dare to believe in the possibility of change— those still caught in the shadows, those who have lost their way, and those who stand ready to lift them back into the light.

So, dive in knowing that this journey will be rough, unpredictable, and at times painfully intimate. But know also that within these pages, there is every reason to hold on to hope, to find strength in vulnerability, and to witness the power of a life reclaimed from the depths. I'm not just telling you the story of addiction, violence, and recovery; I'm asking you to see the person behind the headline, hear the unheard cries, and maybe, just maybe, join me in building a world where redemption is not the exception but the rule. Welcome to ROCKSTAR—where the shadows begin, but the light is never far behind.

Chapter 1
Beginnings in the Crack Epidemic

Childhood Shadows

The early years of my life were etched against a backdrop that was less a nurturing ground and more a battlefield scarred by the relentless waves of the crack epidemic. Our neighborhood, a maze of crumbling brick buildings and graffiti-laced walls, hummed with a constant, tense energy that whispered of despair, survival, and unspoken wars. The streets were narrow and suffocating, lined with rusted-out cars and litter that seemed to pile up overnight, unnoticed by the indifferent gaze of a city that had mostly checked out on us. Every corner was a story of struggle, every stoop a congregation of shadows where dealers moved like ghosts, their transactions brisk and ruthless. The air itself felt heavy, thick with the smell of burnt rubber, sweat, and the acrid tang of chemicals that dissolved dreams before they even had a chance to form.

Our apartment was squeezed into a twelve-story building that had long lost its shine, much like the hopes of those who lived within its peeling walls. It sat on the wrong side of opportunity, cast aside by emerging investments and ignored by the city planners who claimed to have better things to do elsewhere. Inside that cramped space, my family carried on with the daily fight against the odds, trying to patch together moments of normalcy amid chaos. My mother, a woman bruised by life but unbroken in spirit, worked multiple jobs — often clocking late nights in a diner or hustling cleaning shifts in office buildings — while my father's presence was a ghostly flicker, sporadic and unreliable. His own battles with addiction and brushes with the law were family secrets spoken only in whispers and silences. The love in our household was complicated, embroidered with frustration and fatigue, but it was real, nonetheless. Yet, beneath the surface, cracks began to form, mirroring the fractures spreading through our neighborhood.

The crack epidemic was more than just a social crisis; it was a living, breathing entity that consumed everything it touched. It flooded our streets with dealers whose smiles were sharp and cold, their pockets lined with bills stained by the desperation of others. For a child like me, the drug trade was both a terrifying force and a magnetic pull. On one hand, it represented the unvarnished reality of violence and loss — gunfire echoing in the night, neighbors disappearing behind bars or tombstones — but on the other, it was a beckoning escape for those too young to understand the depth of the trap. Early encounters with these streets didn't come out of choice but necessity: the grown-up world around me was fractured by survival, and with survival came the tacit lessons of how to adjudicate pain and hunger.

Even as a child, I sensed the transformative power of the crack epidemic on my community, like a dark tide that reshaped every hill and hollow of our urban landscape. Schools were overcrowded and underfunded, classrooms buzzing with the low hum of despair. Teachers tried their best but were often overwhelmed by the sheer gravity of what their students lugged home each day — not just backpacks, but also the invisible weights of trauma and neglect. The playgrounds, where children should have laughed freely, were places shadowed by the lurking eyes of older kids who carried weapons or knives, their innocence long since corroded by the streets. And so, school hours were a brief intermission from the harshness outside to a place that did little to shelter us from the storm raging beyond its doors.

My mother's warnings about the street were frequent and fierce, delivered with the urgency of someone who had seen too many dreams dissolve into the alleys. "Stay in school, stay away from the corners," she'd say with a heavy heart, but in a world where options were scarce and dangers lurked in every shadow, those words sometimes felt like distant prayers rather than shields. Family dinners were often rushed or missed altogether, the absence of my father growing into a chasm that even my mother's resilience struggled to fill. The siblings I had were both a source of anchor and further complication — older cousins and uncles who flirted with gangs and hustles, younger ones whose wide eyes reflected the uncertainty that nothing good waited just around the corner for us.

In this volatile mix, gangs were as much a part of the scenery as the broken streetlights and boarded-up storefronts. They carved territories with names whispered like prayers or curses, badges of belonging in a world that seemed intent on pushing us apart. The gangs offered something that the neighborhood otherwise lacked: a semblance of protection, identity, and family — albeit one bound by codes of violence and silence. To a young boy craving acceptance and security, the idea of belonging to one of these groups was both alluring and terrifying. It was a promise of immediate respect, a way to shield oneself from the predatory eyes that scoured the streets. But it was also the gateway to a life where trust was currency, and betrayal often meant bloodshed.

The first time I truly witnessed the grim shadow of gang influence was when I was no older than eight. I still remember the chill of that night, the way the streetlights flickered as a black car cruised slowly past the block. Two older boys, maybe fifteen, stepped out, their presence direct and unapologetic, as if the asphalt beneath their feet had sworn allegiance to their cause. They moved with quiet authority, exchanging terse words and glances with the people at the corners. To me, their actions were confusing, neither wholly good nor evil, but a brutal reality that I had to learn to navigate. Weeks later, I overheard hushed conversations in my family about a shooting, neighbors talking about retaliation and lost lives as if it were normal — a grim calendar marking time not by days but by violence.

Amid this landscape, the lure of drugs was omnipresent. From painful firsthand experience, I came to understand how addiction clawed into lives with a grip that was both slow and sudden. My earliest memories of drugs were not about the haze of consumption but about the chaos they sowed within my own walls. My father's absence was marked not only by his physical departures but by the residue of his addiction — the broken promises, the financial holes in our otherwise tight household, the silence that fell like a shroud whenever his name was mentioned. He was a man transformed, fluctuating between warm familiarity and cold estrangement, a shadow flickering at the edges of my childhood. I watched my mother's strength sometimes buckle, caught between despair and fierce hope, as she fought to shield us from the damage wrought by the drug's reach.

Despite the turbulence, there were moments of light within the gloom. Family gatherings with my grandmother, who never allowed the neighborhood's hardships to define us; church on Sundays, where the echoes of hymns briefly lifted spirits; and the rare, stolen afternoons where my siblings and I played tag between piles of debris, laughter peeling through the haze like fragile fireworks. But these moments were brittle, precariously perched in the middle of instability. The neighborhood had a way of reminding us that salvation was often just out of reach, a horizon blurred by the haze of smoke and sirens.

The economic deprivation surrounding us was not just an absence of money but the evaporation of possibilities. Jobs for our parents were scarce and precarious, schools were overcrowded and under-resourced, and community programs were few and far between. The city's indifference was almost palpable, an unspoken rejection that isolated us further and made the social fissures widen. As children, we internalized this abandonment, growing up with the quiet knowledge that the world outside our blocks didn't see us as anything but statistics. The crack epidemic didn't just destroy lives; it eroded the very infrastructure of hope, erasing pathways out and replacing them with cycles of addiction, violence, and incarceration.

Yet, even amid this desolation, there was a kind of raw education — lessons learned not from textbooks but lived through scars and survival. The language of the street, the glance that conveyed warning or invitation, the price of trust in a world where betrayal was currency. These early experiences foreshadowed the winding path I was destined to tread. The neighborhood taught me about vulnerability and toughness, love and loss, desperation and fleeting joy. It was a crucible that forged my identity before I could fully understand its impact, imbuing me with a complex blend of fear and fascination toward the world that surrounded me.

Looking back, those years of childhood shadows were not merely a prologue but the core foundation of my story — a story grounded in the gritty reality of urban survival amid the crack epidemic's relentless surge. The influence of family, the oppressive environment, the early encounters with drugs and gangs, all intertwined to shape the fragile threads of my youth. It was in this crucible that the first cracks appeared in my innocence and the first flickers of resilience took hold. From the depths of despair, the ember of survival was ignited, setting me on a course that would test every fiber of my being in ways I could never have imagined. The neighborhood was both prison and classroom, adversary and unlikely mentor, teaching lessons that would follow me long after I left its broken streets behind.

First Encounters

The summer air hung thick over the cracked sidewalks and broken streetlights of Eastwood, the neighborhood where I grew up—a place where hope was less a tangible thing and more like a whispered myth, told softly behind closed doors before fading into silence. The year was the mid-1980s, but time seemed to stretch and blur here, caught between decades of neglect and the sudden, violent surge brought on by the crack epidemic. The city around us was a fractured mosaic of boarded-up storefronts, graffiti-tagged walls, and the constant hum of distant sirens that never really stopped, only muting and swelling like a heartbeat in the background of every restless night. Poverty was the skeleton on which the community's nerves danced wildly, and the crack epidemic was the poison feeding those nerves—quick, cheap, and deadly.

I remember the days when the streets outside our tenement apartment were alive with the chaotic rhythms of children playing double-dutch and stickball, their laughter mingling with the melodic clatter of basketballs echoing off the brick walls. But even then, lurking beneath that surface of innocence, the shadows were growing longer and darker. By the time I reached eight or nine, the innocent games were starting to feel like temporary reprieves from an encroaching storm. The neighborhood had its own language of survival, and that language was fast becoming saturated with the slang and violence bred by the crack trade. At that age, I didn't fully understand what crack was or the scale of destruction it carried, but I could smell it—faint yet unmistakable—like burning plastic and desperation, curling up from open manhole covers and alleyways where the adults whispered hurried deals and introduced themselves to different shades of danger.

My family was no stranger to struggle. My mother worked two jobs to keep the lights flickering and the fridge stocked, and my father had left when I was too young to remember his face clearly, leaving behind a silence that filled the small apartment with an emptiness no number of meals or blankets could cover. The adults in our building—mostly women working shifts or trying to hold on to any semblance of normalcy—warned us about the changes sweeping through our streets, but those warnings often came too late. When the crack dealers moved in, they brought with them an unnatural calm, a cold order that enforced itself with threats too violent to ignore and agreements too fragile to trust. Windows that once dazzled with cluttered life now stared blankly outward, and stoops that hosted animated chatter turned into hushed meeting points for men in cheap sneakers with hard eyes and hands that danced nervously around small plastic bags.

My first real encounter with gang activity wasn't cinematic or as immediate as I later heard in stories—it was subtle, invasive, a creeping inevitability. One afternoon, as I wandered further from the familiar bounds of my building, lured by the hypnotic sound of amplified music and shouting, I stumbled on a small group of older kids gathered around a burnt-out park bench. Their faces were a mixture of boredom and menace; they seemed to be waiting for something—or maybe someone. The air was thick with the unmistakable aroma of crack cocaine, the sharp chemical tang clashing eerily with the earth and open trash cans around us. For me, that smell will always be the scent of a faded innocence, a smell that marked the point where childhood cracked open and sharp edges began to hurt. The kids eyed me, sizing me up like a prospector sizing up a claim. I was no threat, just a curious outsider from the building next door, but the blue bead of fear that crystallized in my chest convinced me to turn and leave, footsteps echoing loud on the empty street.

What followed were sightings that began to stitch themselves into the fabric of daily life like grotesque murals, impossible to ignore. Drive-by shootings at dusk that turned the sky fluorescent red with gunfire, the wail of ambulance sirens cutting through the thick dusk air, neighbors nervously pulling their doors shut as hooded figures passed by, their movements calculated and impatient. The crack epidemic wasn't just a health crisis—it was the pulse of a violent rhythm that dictated who lived, who died, and who got left behind. The gangs that sprouted like weeds in the blighted streets weren't a single monolith but shifting alliances of desperation, ambition, and survival. Their colors weren't jerseys or simple symbols—they were declarations of identity, claims of territory scribbled in spray paint and gunpowder. For kids like me, the first glimmers of gang influence were not sudden but insidious, appearing in the form of initiation rituals whispered on playgrounds, the exchanged nods between older boys in leather jackets, and the insistent pressure to "know your place" in the social hierarchy that wasn't written down but lived every day with teeth bared.

Inside the dwindling safety of my home, the crisis was less visible but no less pressing. My mother's tired eyes reflected the strain of trying to balance the roles of provider and protector in a world where the stakes seemed impossibly high. Cracks formed in family dinners as news of friends lost to overdose, shootings, or jail made their way into our conversations—each loss a silent wound stitched with sorrow and frustration. The neighborhood's schools were strained under the dual burdens of overcrowding and a curriculum that rarely spoke to the realities of the students it served. Teachers did what they could, but many of us were slipping through the cracks as quickly as our streets were being overrun. The promise of education felt like a fragile boat in a rising tide, and many of my peers began searching for sturdier vessels in the shape of gang allegiance and quick money facilitated by drug transactions.

Even at that young age, the dichotomy was stark and unsettling. On one hand, the streets taught lessons of ruthlessness and mistrust; on the other, whispered stories of escape and redemption fluttered quietly like fragile paper birds, barely able to fly over the smog and despair. I remember watching as older boys came home wearing fresh scars and expanded chests, their confidence forged in relentless violence and blood-soaked initiation. The neighborhood's elders told tales of better days—streets alive with music and block parties, community members looking out for each other instead of watching from behind boarded windows. But those days felt like ghosts, and the crack epidemic had rewritten the rules with an unforgiving ink.

My own introduction to drugs beyond the distant smell and scattered sightings came unbidden and raw. At twelve, curiosity mixed with a confusing cocktail of fear and rebellion led me to an encounter that neither good intentions nor bad warnings could have prepared me for. It was a humid afternoon when a boy from school I barely knew pulled me aside behind the corner bodega, the sun blazing bitterly overhead, and offered me a small rock of crack. I held it with trembling fingers, unaware of the chemical chains tightening invisibly around it, ready to bind me to a pathway filled with shadows that stretched long into the future. The boy's voice was casual but insistent, the kind of tone that made refusal feel like weakness, and the promise behind those few bitter minutes of escape sounded like something impossible to resist when the world around us felt so heavy and cold.

That delicate first taste wasn't just about the drug itself; it was an introduction to a new reality where survival was measured in faster beats, sharper instincts, and a sense of invincibility that both protected and destroyed. Around me, streets that had been backdrops to childhood games were now arenas for turf wars and paranoia. Gang members weren't just threats—they were invisible anchors that tethered you to a life spiraling deeper with each poorly made choice. The crack epidemic fast became the axis upon which our community's fate spun. Families split, friendships shattered, and the neighborhood's fragile pulse flickered in and out with each new tragedy. Police raids and curfews became part of the landscape, but more often than not, law enforcement was just another shadow moving through the streets, unable or unwilling to confront the deeper roots of the crisis.

Even as a kid, it was impossible to ignore how the epidemic echoed beyond crime statistics and headline news; it had a voice and a face in every cracked window, every whispered threat, every desperate plea. It was in my own mother's weary sighs as she locked the door at night, in the vacant eyes of neighbors crushed by addiction, and in the way grown men folded into themselves when confronted by the ghosts of their own decisions. The community's fabric, already frayed by years of systemic neglect and economic collapse, unraveled faster than anyone could stitch it back together, leaving each of us to pick up the threads and either weave new stories or burn what remained.

In the midst of this chaos, I was still just a boy walking a thin line between the worlds—the one inside our apartment, where safety lingered in small acts of care and the promise of family, and the one outside, where the rules were harsher and the stakes much higher. My neighborhood was a classroom of survival, where the lessons were brutal and the curriculum scarred by addiction, violence, and hopelessness. The cracks in the pavement reminded me daily of broken promises, but they were also where I first glimpsed the fierce human will to hold on, to resist, and maybe someday, to heal. From those first encounters with gang activity and drug use, the undercurrent of my life was set—a tumultuous path marked by fear and fascination, destruction and hope, darkness and the faintest, stubborn glimmer of redemption waiting just beyond the shadows.

A Broken Playground

The cracked concrete beneath my feet seemed to mirror the uneasy fissures splitting my neighborhood, a place where the buzz of childhood laughter had long since been replaced by the ominous rhythm of sirens and whispered threats. The playground—the place that was supposed to cradle joy and innocence—had deteriorated into a broken mirage, a warped shadow fighting desperately to hold on to traces of what it once was. I remember standing there, not yet fully understanding how my life, like the neighborhood itself, was unraveling into something unrecognizable. The swings hung crooked, their chains rust-eaten, squeaking in the dry wind, and the faded slides offered no thrill, only a dull invitation to a world I'd no longer be welcome in. It was a wasteland of shattered dreams masked by graffiti-scarred walls and tenement buildings standing like weary sentinels to the havoc creeping in, inch by ruthless inch.

I wasn't born into this hell overnight; the cracks in my world had been forming slowly, exposed by the corrosive acid of poverty and neglect, sapping the very breath from the air we breathed. The crack epidemic had injected itself like a toxic virus, spreading through the veins of my community, turning neighbors into strangers and playgrounds into battlegrounds. Mothers whispered caution to trembling children who were too young to grasp the gravity of their words, while fathers vanished into the fog of addiction or the cold grip of incarceration. What should have been a sanctuary of innocence became a backdrop of constant unease. Gunshots ricocheted through the streets, bones snapped under the weight of violence, and the heavy silence after each eruption bore into your soul like a hammer strike. It was everywhere: the graffiti slogans that twisted from mere vandalism into gang insignias marking territorial dominions, the shattered bottles reflecting the city's fractured promises, and the haunted eyes of kids who had already aged beyond their years.

In that damp and fetid apartment, where peeling wallpaper whispered tales of hardship, my family tried to cling to sanity. My mother's hands, worn and raw from backbreaking jobs, trembled when she tried to hold onto us—all of us, scattered by the storm around us. Her dreams dissolved the moment she realized the streets had swallowed her husbands, one after another, like they were offerings to some dark god demanding sacrifice. My sister's quiet sobs and my father's distant stares were the soundtrack of loss reverberating through the cramped rooms. We were a family fractured along invisible fault lines, not broken by choice, but by the relentless pressures of a world that had forgotten us. The promises of the American dream did not extend to us; they dangled like cruel mirages, just out of reach while the reality of hunger, fear, and desperation pinched tighter with each passing day.

Walking the worn sidewalks, I was keenly aware of the razor-thin line separating me from falling deeper into a vortex of despair. The neighborhood wasn't just an environment—it was a living, breathing entity, pulsing with danger and possibility, enticing and repelling all at once. The older kids, hardened by necessity, were already leaders in street corners where the drug trade blossomed like a poisonous flower, offering a twisted form of survival to those willing to risk it all. Their eyes, hollowed by addiction and violence, bore down on me with a mixture of challenge and invitation, whispering promises of respect, power, and escape from the grinding monotony of poverty. For some, the lure was irresistible—the cold commerce of crack delivered quick relief and intoxicating status, but it came with shackles heavier than chains. I remember watching the boys I used to play with transform almost overnight into ghosts of themselves; smiles vanished, replaced by furtive glances and twitchy hands.

My earliest encounters with drugs weren't steeped in rebellion or curiosity as much as they were in survival—a desperate grasp for warmth in a cold world. It was whispered rumors and furtive exchanges behind rusted dumpsters, the clinking of glass pipes passed between trembling fingers, and a false sense of brotherhood forged in shared desperation. There was no glamour in it, only a grim rhythm dictated by the rules of the street. The gang influence was a shadow lurking in the corners of every block, its grip tightening as the epidemic swept through like a tidal wave. It marked us, identifying friend and foe, weaving a complex tapestry of allegiances and betrayals that shaped our identities and futures before we even had a chance to understand ourselves.

At school, the halls were filled less with the innocent chatter of friends and more with dark talks and silent fears. The teachers, overwhelmed and under-equipped, did their best to reach a generation that felt more connected to the street's code than to any spoon-fed curriculum. I sometimes wondered if they really saw the kids behind the classroom desks—the way our eyes darted constantly, searching not for knowledge, but for safety. The halls echoed with the weight of whispered threats and the gathering storm of growing gangs fighting for control of turf that defined survival more than geography. Innocence was a currency already spent, traded for scraps of protection and fleeting moments of belonging in a world fiercely indifferent to our plight.

These early years were a constant battle against a creeping numbness, a desperate plea for meaning amid the chaos. Every shattered window, every siren that split the night, every loss felt compounded the erosion of a childhood I barely recognized as mine. I stood at the crossroads of choice and circumstance, my small hands trembling as I reached out for something solid, something that would keep me from slipping entirely into the void. But all around, the broken playground held only shadows, echoes of a life lost before it had truly begun. That loss—the slow death of innocence—was the first wound, heavy and unseen, setting the stage for everything that was to come.

Chapter 2
Temptations and Traps

The Gang's Call

The summer sun hung low like a dull ember over cracked pavement and broken dreams, casting long shadows across the worn stoops and graffiti-scarred walls that formed the backdrop of my teenage years. The air was thick with the scent of sweat, burnt rubber, and something darker—an almost electric buzz that mingled with the distant sirens and the low rumble of cars rolling slow, windows down, music blaring old hip-hop anthems that spoke of struggle and survival. It was in this raw, unforgiving terrain that the gang's call first echoed in my ears, a seductive siren song that promised blood, brotherhood, and a way out of the suffocating grasp of poverty and insignificance. It came cloaked in camaraderie, whispered in the corners of broken playgrounds and alleys stained with forgotten fights and unspoken regrets, pulling me deeper into a world where loyalty was currency, and survival came at the edge of a blade.

I was barely fifteen when the invitation came—not on paper or through any formal rite, but through gazes lingering too long, through words thick with menace and affection wrapped around the same breath. It was an initiation not just into a crew or a clique but into a mindset, a way of being that offered belonging in the absence of family warmth, a release from the relentless boredom and bleakness of day-to-day existence. I had always been a kid straddling two worlds—the school corridors where teachers barely looked twice, and the streets where respect was earned through fear and power. It was on the streets, though, that the gang's pull grew irresistible, a gravitational force made stronger by desperation and loneliness. I remember the way their members moved—confident, ruthless, speaking a language of toughness and silent codes that I craved to be part of, even if I didn't fully grasp the cost at the time.

Peer pressure wrapped around me like a suffocating cloak, every interaction a subtle test of allegiance. Refusal wasn't really an option when the alternative was invisibility or worse—targeted for being weak, or worse yet, seen as an enemy. The initiation came not as a one-time event but as a series of small, escalating steps. The first time I was asked to run an errand, I was scared but eager. I recall clutching a small plastic bag with contents I barely understood—powdered substances packed tightly and folded into the seams—my heart thudding wildly against my ribs as I walked streets that suddenly felt both familiar and alien. Each errand I completed stitched me tighter into the gang's fabric, a thread pulled taut with tension, excitement, and fear. The act of carrying those small packages was not just a test of trust but a baptism into a harsh new reality where the stakes were life and death.

The gang's promises glittered with a cruel brilliance. They spoke of power, protection, and respect—the kind that schoolteachers and social workers never handed out. For a kid growing up in the shadows of crackpipes and gunshots, those promises shone like beacons, a false holy light in a world already dark. With each visit to the corner, each whispered plan behind chain-link fences, the allure deepened. They promised legacies beyond the broken home I came from, a replacement family where you had a name, where your presence demanded attention. The idea of sitting silently at the dinner table, ignored and overlooked, became intolerable compared to the thunderous roar of the streets that now beckoned me. The gang wasn't just about crime or drugs—it was about identity. It was a tribal call that answered an empty chamber inside me that the suburban dreams of college and white picket fences couldn't reach.

Yet, the dangers lurked just beneath the surface of those promises, like shadows waiting to pounce the moment you blinked. The initiation was a gauntlet lined with risks masked as rites of passage. I saw friends fade into the night, swallowed by gunfire or jail bars, their lives reduced to cautionary tales whispered around the fire. I watched trusted comrades bleed and vanish, swallowed whole by the shadows we once chased with youthful bravado. The gang taught me the brutal lesson that trust was a fragile thing, quickly fractured by betrayal or a misstep. In that world, once you stepped inside, you were no longer a boy with dreams but a soldier in an unending turf war—a pawn caught between desperation and desire. Each initiation meant one more step away from innocence and one more rung climbed on an unstable ladder that could snap at any moment.

My own initiation culminated one humid night beneath the flicker of streetlights that barely cut through the thick fog of smoke and tension. It wasn't flashy or cinematic; it was jagged, raw, and terrifying. I was handed a gun for the first time, heavy and foreign in my trembling hands, the metal cold but somehow pulsing with a dark life of its own. There was no grand ceremony, just a curt nod from a leader whose eyes told stories of battles won and lives lost. The instruction came terse, almost mechanical—respect the gun, trust your instincts, and above all, watch your back. That night, I stood amidst a circle of faces painted with shadows and anger as voices around me murmured the unspoken creed of the streets—the pledge of loyalty sworn not with words but with the readiness to spill blood or be spilled. Fear mixed with adrenaline in a cocktail that quickened my heartbeat and dulled my senses. I realized then the full weight of the gang's call: it demanded not just participation but sacrifice, not just loyalty but surrender to a force larger than any one boy.

With the gun came substances—easy, potent, and deadly temptations that promised escape from the crushing pressure and gnawing doubt. The first time I smoked crack, it wasn't in a grand rebellion but in furtive inhalations shared with brothers who claimed it sharpened courage and numbed pain. The crack epidemic had seeped deeply into our neighborhood like a creeping illness, infecting hopes and laying waste to futures. It offered moments of euphoria, brief respites from the reality of bullet-riddled blocks and desperate nights, but beneath the glittering smoke was a poison that slowly eroded everything it touched. For me, drug use began as a means to fit in, a response to unbearable stress and fear, but it quickly spiraled into dependence—a shadow companion on the long nights filled with paranoia and regret. Each hit etched lines of destruction deeper into my soul, blurring the boundaries between freedom and imprisonment.

The gang taught me how to talk, how to move, how to survive. It taught me that weakness was a liability and that pain was something to be swallowed, hidden behind a mask of toughness. But it also taught me about loss—the way friends could vanish overnight, their names whispered with reverence or fear, their faces frozen in memories sharper than knives. I lost many of my own in that crucible—some to bullets, some to the dark pull of addiction that strangled hope with its invisible hands. Each loss imprinted a lesson carved into my heart; each scar, physical or emotional, was a testimony to the perilous path I had chosen. Yet the gang's call was a siren I couldn't resist, a code written in blood that promised meaning amidst chaos, even as it slowly consumed the boy I once was.

As my teenage years unfolded in this relentless storm, the boundaries between right and wrong, safety and danger, blurred into a single, chaotic line that I stumbled along every day. The streets became both my prison and my refuge, offering the twisted duality of belonging and destruction. The gang's call echoed in every decision, shaping my identity, my choices, and my future in ways I could barely comprehend. I learned to navigate the dangerous dance of power and fear, weaving through moments of camaraderie and conflict, love and betrayal. What began as an invitation to belong became a demand to fight, to endure, to survive at all costs. The shimmer of promised respect came paired with the harsh reality that each step further into the gang's world took me deeper into a darkness that threatened to swallow me whole.

Looking back, the gang's call was not just an invitation, but a trap disguised as hope. It was a seductive force that promised escape but delivered chains. It pulled me into a violent, addictive world where every day was a gamble with death, and every moment's peace was paid for in fear. It was a rite of passage stamped in bullets and crack smoke, a brutal education in the costs of survival on the unforgiving streets. Yet, even in the depths of that world, the human heart reached for something beyond the violence—a flicker of hope, a chance at redemption, a life reclaimed from the shadows. The gang's call became the crucible in which I was forged, a painful beginning to a story of fall and, much later, rise.

Money and Power

Money and power were seductive aphrodisiacs that wrapped tightly around my teenage soul, pulling me deeper into a world I once only glanced at from the edges. It began in flickers—small bursts of cash that broke through the fog of poverty that clung to my family like a second skin. I remember the first time I held a fat roll of bills, the ink still fresh, the paper crisp beneath my calloused fingers. It was not just money; it was a ticket out of the constant hunger, the stifling heat of cramped, damp apartments, and the endless cycle of watching grown-ups drown in despair and addiction. That cold, heavy stack of bills whispered promises of control, respect, and survival in a neighborhood where those things were currency just as valuable as any green paper.

The alleyways and cracked sidewalks of my block morphed beneath my feet into a labyrinth where power grew like an invasive vine—tough, thorny, yet intoxicating. As a kid, power had always been an abstract, distant concept. It was in the way certain older guys could command a room with just a glare or a sharp word, the way their presence made even the bullies pause. It was in the propulsion of a quick nod or a handshake that could open doors to a quarter bag of powder or a stash of cash in bulging pockets. And suddenly, I wanted in—not just to survive, but to thrive amidst the chaos, to carve a space where I belonged and could dictate my terms.

Peer pressure hammered at my door relentlessly, splitting my world into the binaries of those who played the game and those left behind to fade into obscurity. My friends, eager and equally lost, became partners in crime, bonded by the magnetic pull of fast money and fleeting respect. We weren't just kids messing around; we were soldiers in an underground war, and every dollar earned was a badge of honor. The street slang rolled off our tongues like a second language, and with every hurried exchange in dimly lit corners, the thrill of control surged—like a shot fired straight into the heart's nerve.

I can still feel the heat of those late nights when the city felt alive in a dangerous pulse, neon signs flickering against graffiti-stained walls, sirens wailing distantly but never close enough to stop us. The deals were transactions not just of drugs and cash but of trust, fear, and allegiance. The money stacked high in my pocket felt like armor. Suddenly, I had a voice that was heard, a presence that was feared—a far cry from the boy who once hid from his own shadow. That pretense of control was intoxicating, a high more addictive than any substance at first. I felt untouchable, as if money alone could rewrite my story, change the narrative of deprivation I'd been born into.

Yet, beneath that false veneer of power lurked the undercurrent of danger and desperation that I was slow to recognize. The deeper I waded into the current of gang life, the more it shaped my sense of identity until it became inseparable from who I thought I was. I was armored in bravado, but inside, the gravity of each decision weighed heavily. Every deal struck was layered with risk—a wrong word, a twitch of suspicion, the glare of law enforcement around the corner. But the promise of power disguised the growing cracks in that armor, blinding me to how fast the world could turn from invincibility to nightmare.

My teenage years, already turbulent with the raw and reckless energy of adolescence, became a battleground where loyalty and survival collided. The pressure from my peers was relentless, a chorus that demanded conformity with the code of the street. To step back meant certain ostracization, vulnerability, and an erasure from the community that had become both home and prison. My choices were no longer solitary acts but parts of a collective that demanded obedience, that rewarded aggression, and punished hesitation with brutal finality. The initiations were not just rites of passage—they were thresholds to a life where money and power were both the prize and the prison bars.

As I embraced this lifestyle, substance abuse crept in as an insidious companion. The drugs weren't just commodities—they were tools that numbed the sharp edges of fear and guilt. The crack I sold, once just a means to an end, soon seeped into my veins, blurring lines between dealer and user. The more control I thought I had, the more the addiction tightened its grip. Late nights morphed into hazy blurs; days bled into nights, and the chase for the next hit began to overshadow the chase for money. Yet paradoxically, that initial taste of power stayed with me, haunting me like a mirage in the desert, sparking a fierce stubbornness to hold on to what little agency I felt I possessed.

In those moments, I was caught in a whirlwind of contradictions. The money gave me access—to food, clothes, fleeting luxuries that felt like a throne carved out of my struggles—but it also made me a target. Friends turned into foes overnight; alliances forged on shaky ground cracked under the weight of suspicion and greed. Violence was the language of survival, an unspoken rule that made every transaction a gamble, every corner a potential battlefield. The adrenaline rush of power, of being feared and respected, was intoxicating yet terrifyingly fragile.

Looking back, I see how these early profits were not just currency but catalysts, fueling a cycle of deeper entanglement. Each dollar earned was a thread in a tangled web that ensnared body and soul. The sense of control was an illusion, a brief flicker of light in a landscape that increasingly grew dark and oppressive. And yet, for a long time, that illusion was enough. It was a lifeline, a semblance of dignity in a world that had long denied me both. The money whispered possibilities, the power promised escape, and I leaned into both with reckless abandon, unaware that the very things I chased would become the chains that bound me.

Saving some of those early earnings felt like staking a claim to an empire I was building on fragile sand. I bought things—a jacket that made me look bigger, tougher, someone not to be trifled with; a pair of sneakers that gleamed under streetlights like trophies. These tokens of success weren't just material; they were affirmations of a self forged in the fire of desperation and desire. I remember the way neighbors would look at me differently, some with envy, others with a wary respect that whispered, "He's made it this far." But the deeper truth was that the money was a double-edged sword, tightening the noose even as it offered a crown.

The streets taught me brutal lessons in the economy of fear and loyalty, where money was both shield and weapon, a language spoken in nods and glances rather than words. The dopamine hit of a quick sale, the weight of cash exchanged in hushed voices, became addictive rhythms that drowned out the emptiness that loomed beneath. Even as the profits grew, the quiet moments revealed a gnawing hollowness—a realization that this power was borrowed, conditional, and fleeting.

In those years, every choice was etched with urgency, with a desperate grasp for control in a life spinning out of hand. I was hungry for respect, yearning to prove I could rise above the fractured circumstances of my childhood. But I failed to see that the money and power I craved were fleeting dreams built on shifting sands of danger and deception. The sense of mastery I felt was only skin deep, barely shielding the fractures that would soon crack wide open.

And yet, in that raw, chaotic crucible, I learned something vital: that power, when divorced from purpose and grounded only in survival, becomes its own prison. The money bought momentary freedom but never lasting peace. It was a siren's call, a promise turned curse, and I was caught in its deadly embrace—blind, hungry, desperate for meaning in a world that offered so little. The dollar bills in my hand were like shattered mirror fragments, reflecting a fractured self trying to piece together an identity amid the smoke and fire of a world descending into darkness.

Crossing Lines

The evening air hung heavy with a damp heat that clung to my skin like a second layer of grime, a familiar sensation that followed me through the narrow corridors of the neighborhood. Cracked concrete and flickering streetlights set the stage for a scene I knew all too well — the restless murmur of groups huddled beneath graffiti-stained walls, the faint echo of distant sirens slicing through the night, the occasional crack-snapping burst of laughter that seemed hollow given the weight pressing down on us. That's when the lines started blurring — those clear but fragile boundaries between right and wrong, lawful and unlawful, safety and danger — all folding into one indistinguishable gray zone where choices felt less like decisions and more like survival tactics ingrained by circumstance.

I was sixteen, but the world I moved through felt older, harder, and more unforgiving than the streets would allow any kid to be. My life revolved around the corner store, the parks where we gathered, and the social hierarchy of gangs that dictated the flow of power and respect. The transition from boyhood to manhood was not marked by celebrations or milestones but by acts that chipped away at whatever innocence was left. Peer pressure wasn't just words thrown around; it was a gnawing force that shaped everything I did, pressed into my skin and bones with the weight of expectation and fear. The friends I looked up to, the older crew members whose accolades came from their daring, their ruthlessness, their willingness to break every rule, became the standard I subconsciously chased. I wanted to belong, to be seen as one of them — not just for the false safety of numbers, but for the illusion of power they wielded, the kind I thought would shield me from the chaos at home and in the streets.

The first time someone handed me a joint, it was almost ceremonious, as if passing off a torch of adulthood. It wasn't a choice born from curiosity or rebellion but rather a script written by those around me — a ritual that sealed a fragile bond of acceptance. Smoking offered a brief escape, a momentary drift into a haze where the raw edges of the world blurred into something bearable. Yet as the nights passed, what started as casual clouds of smoke evolved into heavier substances, whispers of crack that promised a sunburst of euphoria but delivered a descent into shadows darker than any streetlight could illuminate. The crack epidemic was not an abstract crisis; it was a creeping menace wrapped in dollar bills and desperation, spreading its poison through blocks and families like wildfire. To say I was drawn to it is almost a disservice — it was a tidal pull, relentless and inescapable, dragging me further down a path loaded with peril and regret.

Risk was a currency exchanged with every hurried deal and every glance over the shoulder. The neighborhood had rules, unspoken codes enforced with violence as a universal language. I remember the first time I was asked to carry a weapon — not as a joke, not as a token of trust, but as a necessary tool in what we falsely thought was self-defense. A rusty, chipped pistol slipped into my jacket with the cold firmness of inevitability. The weight in my pocket was more than steel and metal; it was the burden of stepping into a role I never sought but was thrust upon me. There was no fanfare, only a silent understanding that crossing this threshold meant leaving behind the small cracks of childhood I had tried to hold on to. That night, as I lay awake with the gun beneath my pillow, I felt an anxious thrill mingling with a creeping dread — a cocktail of emotions that would come to define so many nights after.

Gang life, once a spectacle from a distance, became the beating heart of my existence. Loyalty demanded blind allegiance, and trust was rationed like precious commodities. Every morning was a calculation — whom to avoid, whom to watch, and whom to fight. Conflicts flared over trivial matters but ignited infernos that left scars on bodies and psyches. I saw friends fall like dominoes, taken down by rival crews or by the unforgiving reach of law enforcement. The promises of the streets — respect, money, status — tasted bitter and hollow in the aftermath of each violent confrontation. Yet the hunger for belonging and meaning kept us entangled in cycles of retaliation and survival, perpetuating a narrative where every act carried the weight of future consequences.

I recall one particular night, the memory seared like an indelible mark, where the stakes came crashing into clarity. It started as a whispered quarrel over a misunderstanding — some insult or perceived betrayal blown out of proportion by jittery nerves and bruised egos. Words turned to shoves, shoves to fights, and soon, blaring sirens wove their way to the scene as chaos erupted on cracked asphalt. I found myself caught in the crossfire, fists swinging, heart pounding like a drumbeat in an endless rhythm. The adrenaline was a cruel accomplice, clouding judgment and fueling aggression, eroding any lingering restraint. In that moment, rules dissolved entirely, replaced by primal urges to fight, protect, and survive. When the dust settled, someone lay bleeding at my feet — a crossing of a line neither of us wanted but had stumbled into without realizing.

That incident was pivotal, a flashpoint revealing just how easily moral and legal boundaries could shatter under pressure. The sense of invincibility I clung to was a fragile illusion. From then on, consequences loomed larger, and the shadows lengthened. Yet the pull of the streets was magnetic, fueled by fear of exclusion and a desperate need to assert control in a life governed by chaos. I felt trapped between aspirations and impulses, between the flickers of hope and the suffocating grip of addiction and allegiance. This double bind was a relentless loop — each choice narrowing the corridor toward darker alleyways and deeper setbacks.

With each passing day, my injury to the self became more apparent — cracks splintering a foundation already shaky. The substance-use spiraled — crack revealing itself not only as an enthralling escape but as a merciless master, demanding ever more in exchange for fleeting relief. Sleepless nights bled into blurred mornings, and the vibrancy of youth was dulled into a monochrome existence marked by paranoia and cravings. I watched as friends faded into ghosts, swallowed by overdoses or locked behind iron bars. Every face in the neighborhood bore stories of loss; yet, the cycle persisted, fueled by systemic neglect and the crushing weight of poverty.

The law was no haven. Early arrests — small, inconsequential to many — began to accumulate, tattoos of failure etched in police records and courtroom hearings. The courthouse smelled of stale coffee and forgotten dreams, a place where justice often felt uneven, falling harder on those trapped in desperation than on the forces that ensnared us. The looming threat of incarceration shadowed each decision, a constant reminder that crossing the line wasn't a one-time event but a series of sinking steps deeper into a quicksand of consequences. Youthful bravado masked the growing fear, yet each interaction with the law peeled away layers of invulnerability, replacing them with a cautiousness sharpened by experience.

But the lines weren't just external. Inside, I grappled with a moral warfare — flashes of conscience wrestling with the ruthless demands of survival. I questioned the path, sometimes in the quiet moments when the streets retreated, and loneliness crept in, creeping past the bravado to expose cracks in my resolve. The reflections came sharp and sudden, like winter's chill slicing through a summer night. I saw the faces of family members, those who loved me despite my flaws, and felt the tug of something more human than the hardened shell I wore. Yet the gravity of addiction dulled these moments, pulling me back down with a force that was both physical and emotional.

Peer pressure was relentless, its tendrils tightening around my sense of identity. If you didn't move with the pack, you risked becoming prey. Standing apart was dangerous, equated with weakness or betrayal. I learned to mask fear with aggression, to speak the vernacular of the streets with an ease that concealed the turmoil beneath. Being part of a crew was armor, but it was also a cage, constraining futures and aspirations into rigid molds forged by desperation and limited options. The pursuit of respect often came at the cost of self-respect, and the need to survive took precedence over dreams deferred.

There were rare bursts of clarity amid the fog — moments when I glimpsed the possibility of a life unshackled from this revolving door of violence and addiction. But these flashes were fragile, often extinguished by the next confrontation, the next hit, the next betrayal. The tension between hope and despair was a constant companion, a tug-of-war that shaped each day and night. The streets whispered a seductive but dangerous lullaby, promising belonging and power but delivering scars unseen and wounds unhealed.

In those teenage years, crossing lines became not just about breaking laws or codes but about slipping into a new identity, one forged in fire and frost. Each step deeper into gang activity and substance abuse was a fissure widening, a fragmenting of the person I once was and could have been. The cost was steep — innocence lost in smoke and blood, futures traded for borrowed moments of power, and a soul caught between the shadows and a distant, flickering light. This was the beginning of a long, harrowing journey through darkness, where every choice marked another crossing, another step away from safety and toward an uncertain redemption that I could only faintly imagine.

The High Cost

The teenage years were a murky haze, a relentless collision of choices and consequences that I could never quite outrun. It was in those early days, when the world still felt vast, and the streets whispered of endless possibilities, that I first felt the creeping weight of the high cost—though back then, I didn't dare give it a name. I only knew the tight grip of pressure, the sharp sting of expectation, not just from family but from the shadows that lurked in the alleys where I spent my afternoons and late nights. The crack epidemic was swallowing my neighborhood whole, like a beast carving deep furrows into the pavement, stripping away whatever innocence remained. I remember the scuttle of feet, the quick glances behind corners, the slang tossed like daggers between hardened boys who were only a few years older than me, yet seemed forged from entirely different steel. They were the gatekeepers, the currency dealers, the enforcers of a brutal economy built on desperation and fleeting highs. Their world was glamorous on the surface, shimmering with the promise of quick money and respect—the elusive currency of survival when the traditional routes seemed barricaded by poverty and systemic neglect. But beneath that surface lay a relentless erosion of everything that mattered: trust, safety, hope.

Each day felt like walking a taut wire, balancing between temptation and resistance, though the wire frayed under the weight of peer pressure. I was a young kid, clutching feverishly to whatever scraps of control I could find, listening to the sirens wail in the night while the chatter of dealers and gang members hummed through the cracked walls of my home. It was as if the city itself was a breathing entity, pulsing with the adrenaline of violence and the lure of escape through substance. Some nights, when the smoke thickened and the shadows grew long, I caught glimpses of what life could be like outside the confines of my block—a ghost of a distant dream where laughter wasn't punctuated by gunshots and futures weren't traded for a bag of white powder. But those moments slipped through my fingers like the ashes of a joint, too fragile to grasp, drowned under the roar of survival needs and the insidious normalization of chaos.

The first time I said yes to crack was not a moment of triumphant rebellion—it was a quiet surrender, a desperate bid for relief from the weight pressing down on my chest. I was barely fifteen, trembling with a mixture of fear and reckless curiosity as a peer pressed the pipe into my trembling hands. There was no gentle initiation, no shimmering gateway to a better world; just a harsh inhale followed by a rush that ripped through my veins, cruel in its intensity but seductive in the illusion of escape it offered. I tasted freedom for the first time, a freedom that was as temporary as it was destructive. And as the high ebbed, reality slammed back in harder than before, the streets calling me deeper with voices loud and unforgiving. Soon, the pipe became a companion, a crutch that silenced the screaming void that grew every day within me. Addiction began to swallow my identity piece by piece, reshaping me into someone I barely recognized—a ghost drifting through the remnants of my fractured neighborhood.

Around me, the gang's influence seeped through every crack in my life. It wasn't just a group; it was a force of nature—born from neglect, fed by pain, and sustained through violence. The initiation rites were brutal lessons in loyalty, pain, and power. Blood spilled not just on the pavement but in the quiet corners of homes, fracturing under the pressure. Each scrape, each fight, each tactical alliance etched a deeper mark on my soul. The gang became an unyielding family, offering the kind of acceptance I craved but could never find in the schools or houses where hope had long since evaporated. But that acceptance demanded a harsh price: that I bend, break, and sometimes snap the humanity inside me into something unrecognizable. I learned early that walking away wasn't just difficult—it was almost impossible. The web of fear, pride, and survival wrapped tightly around me, cutting off alternatives before I could even glimpse them.

Violence became the soundtrack to my existence. The air was thick with tension, every corner a potential battleground. Gunshots echoed in the distance like grim reminders that life was fragile, a currency to be risked and often lost in the blink of an eye. I came to know the faces of grief too well—the friends who never made it past eighteen, the mothers who cradled lifeless bodies, the fathers buried in silence. Blood stained our streets, painted a grotesque mural of loss that weighed heavily on anyone who dared to look too close. Still, we marched forward, numbed to the carnage because there was no space left for mourning. Survival demanded numbness. We learned to carry anger like armor and despair like a secret wound hidden beneath bravado. Each violent episode wasn't just a headline or a statistic; it was a concrete matter that shaped our bones and hardened our hearts. The guilt and sorrow flickered beneath the surface, but they were buried deep—too dangerous to surface lest they break the fragile order holding our worlds together.

At the same time, the community around me slowly crumbled, a tragic domino effect set in motion by addiction and violence. Local businesses shuttered, storefronts boarded up with thick planks of wood like protective shields. Schools dwindled in resources and spirit, teachers abandoned by bureaucracy, students lost in the shadows of neglect. Families fractured under the strain of addiction, hope diluted by the corrosive presence of drugs that invaded homes and hearts alike. The police presence, meant to safeguard, often felt more like an occupying force—alienating rather than protecting, intensifying the cycle of suspicion and resentment. The street itself became a battleground for power and survival, caught between law enforcement and gang allegiances. All of this wove together to create a tapestry threaded with despair, confusion, and a yearning for something different, something better.

In those years, I realized something profound, though I was only beginning to comprehend its full weight: addiction and violence weren't just personal demons—they were systemic monsters, fed by poverty, neglect, and broken promises. They didn't choose their victims randomly. They thrived on the failures of a society that turned its back on neighborhoods like mine, that allowed pain to fester in silence rather than confront it head-on. I saw how the cost stretched far beyond my own bruised skin—it bled into every corner of my community, draining strength and potential, silencing dreams. I watched mothers weep for sons caught in cycles of violence, children robbed of their childhood by the drugs that shadowed every street corner, neighbors who'd once smiled and helped each other now staring empty-eyed through the veil of addiction. It was a slow, grinding destruction, more devastating in its invisibility than any bullet wound—a collective wound that refused to heal.

Yet even in the darkest moments, a faint ember flickered inside me—a mixture of anger, pain, and the desperate hope that maybe, just maybe, this cycle could break. That ember burned quietly while chaos roared around me, a fragile whisper reminding me there was a price to pay for every high, every violent act, and every shortcut to survival. The high cost was not just the loss of years or the fracture of blood ties; it was the soul of a neighborhood, the dreams of children growing up too fast, the memories of lives once full of promise now trapped behind barred windows or beneath cold earth. And it was that realization—raw and unyielding—that would one day spark a yearning to reclaim, to rise from the wreckage, and ultimately to seek redemption. But in those teenage years, caught between the pulse of street life and the fading glow of boyhood, that future felt like a distant star—shimmering, yes, but so far away it barely seemed real. The paths I chose then were marked by shadows, by the relentless pull of addiction and the twisted bonds of gang loyalty, a landscape painted in smoke and blood from which escape seemed a fantasy. Still, the high cost was etched deeply into my bones, a lesson harsh and unforgiving, but one whose meaning I refused to let slip away, even if I couldn't quite understand it yet.

Chapter 3
The Spiral Deepens

Addiction's Grip

The first time the crack hit my lungs, it was like a bolt of lightning cutting through a dull, aching fog that had settled deep inside me. The world sharpened instantly—a jagged, blinding clarity that both terrified and seduced me. In those moments, the weight of everything pressed upon me—the poverty, the violence, the endless desperation—lifted as if suspended by some cruelly delicate string. But that high carved its own prison, tightening around my chest, squeezing the life from me slowly, relentlessly. Addiction didn't creep in as a gentle tide; it stormed like an unforgiving hurricane, uprooting every semblance of normalcy I had left. When I think back now, what stands out most isn't the glamorous illusion I chased but the beginning of the suffocating grip that tightened with every hit, every hit that came faster than the last, pulling me deeper into a spiral I barely recognized until it consumed me.

The physical effects were immediate and invasive, hijacking my body without mercy. My hands trembled unnaturally, quaking like brittle leaves caught in an unending gust. My heart thumped violently, like a war drum hammering an erratic beat against my ribs, threatening to burst free. Sleep became a stranger whose face I could no longer recall; nights stretched interminably as I chased the next fix to push back the yawning shadows that lurked behind my eyes. My skin grew rough and clammy, my face hollowed, and my breath carried a sharp, chemical bite. These were not subtle transformations but brutal markers of the war waging inside me, each physical betrayal a testament to the crack's ruthless domination. At times, my body would revolt violently—nausea tearing through my gut, cold sweats drenching my sheets, and hallucinations flickering at the edges of my vision, dark phantoms born from withdrawal's cruel hand.

But the psychological grip was far more insidious, weaving itself into the fabric of my thoughts, reshaping my mind to serve only one purpose: to feed the addiction. Rationality slipped away like water through cracked fingers, leaving behind a frenzied core of craving that eclipsed every other feeling. I found myself caught in endless cycles of obsession, the crack's siren song echoing louder with every heartbeat. Fear and paranoia bled into every corner, twisting neighbors' glances into threats, turning the shadows into conspirators. The simplest interactions became laced with suspicion, and trust evaporated as quickly as the high. Moments of clarity were brief and harsh, slamming into me like cold water, revealing how tethered I had become to a poison that both numbed and destroyed. And yet, the thought of letting go felt like staring into a bottomless abyss, a chasm of pain and emptiness I was too afraid to face sober.

The emotional devastation was no less profound. I watched, almost with a detached numbness at first, as ties that had once been strong unraveled before my eyes. Friendships fractured under the weight of my erratic behavior and desperate lies; promises made in fleeting moments of lucidity were shattered by betrayals I hardly had the strength to own up to. My family, those fragile anchors in the turbulent sea of my existence, bore witness to my unraveling with a mixture of heartbreak, anger, and helplessness. Phone calls fell unanswered, visits grew scarce, and messages turned cold or vanished altogether. Even when I reached out—rarely, but sometimes in the bitter aftermath of a spiral—they met my words with weary skepticism or outright rejection. The stigma of addiction, the fear of the disease's contagious pull, created barriers that felt impenetrable. Loneliness deepened even as the drug promised companionship, offering hollow solace while stripping away every genuine connection I once had.

As my dependence tightened, interactions with law enforcement became an almost routine menace, punctuating the chaos of my existence with cold, unforgiving reality checks. Police sirens wailing in the distance were the soundtrack to nights spent pacing the streets, heart pounding not just from withdrawal but from dread. The city's cracked sidewalks and alleyways, I once navigated with a tentative sense of control, turned into traps, each corner a shadow, a potential threat, each glance a possible arrest. Arrests came like dark milestones: handcuffs biting into wrists, the sterile interrogation rooms where questions and accusations blurred together, the harsh clank of prison gates shutting behind me. Jails offered no refuge, only stark reminders of the consequences I'd run from but never escaped. Behind bars, the illusion of power dissolved completely, replaced by the raw, grinding humiliation of dependence exposed and punished. But even these moments of enforced sobriety were fragile; the cravings lurked behind steel bars, waiting to reclaim their hold at the first taste of freedom.

I remember one night in particular, the night my addiction fully revealed itself not just as an enemy but a tyrant overrunning every shard of my will. I was seated on a cracked stoop, the city drowning in rain, the wet streets reflecting the buzzing neon like broken mirrors. My hands shook uncontrollably, sweat mingling with the cold droplets sliding down my face. My mind raced between dizzying highs of craving and the crushing lows of withdrawal, a beast locked in endless battle. There was no way out—or so it seemed. I barely noticed the red and blue lights slicing through the darkness until the heavy hand of a police officer gripped my arm, hauling me into a cold, unwelcome reality. That night was more than a legal detour; it was a mirror held up to my face, brutal and unyielding, showing me the abyss I was hurtling toward with eyes wide shut.

With each encounter, the pattern repeated itself—initial defiance, followed by bitter resignation. I clung stubbornly to the drug as if it were the last thread holding me together in a world that seemed hellbent on tearing me apart. It whispered lies in my ear, promising strength, escape, and control, all while tightening the chains that enslaved me. The disparity between who I was underneath and the cracked-out specter I'd become widened daily, a chasm filled with self-loathing and grief. I wanted to claw my way back, to breathe air free from chemical bondage, but the ways back were obscured by a thick fog of shame and fear. Each failed attempt to quit was a sucker punch, each relapse a fresh wound reopening scars I'd tried to hide.

Friends who hadn't already vanished tried to pull me from the edge, but I pushed them away with venomous words or by simply disappearing into the dark corners where dealers lurked. The crack community was a twisted family of sorts—a mixture of desperation, loyalty, and shared destruction. Some were victims like me, others predators eager to exploit weakness. Trust here was a currency rarer than cash, and betrayal lurked as surely as hunger. Nights hazy with smoke, voices slurring promises no one intended to keep, faces twisted in misery and rage; these memories lingered like ghosts, a grim parade of the people I wished I could forget but couldn't unless I freed myself. Too many times, I watched brothers fall, collateral damage in a war waged not by armies but by greed and neglect.

Yet, amid the devastation, there were fleeting moments of stark clarity—precious seconds when the fog thinned, and I could see the wreckage of what I was becoming. These glimpses were painful and paradoxically filled with both despair and hope. I realized that the drug had not taken just my health or my freedom, but my identity. I was becoming a shadow, a caricature of the boy who once dreamed of something better. In this abyss, I grappled with the darkest parts of myself, confronting fears and failures long buried beneath the layers of addiction. The internal battle was exhausting, a ceaseless tug-of-war between self-destruction and survival, where every choice felt like a narrow escape or a fatal fall.

This duality—the intoxicating euphoria that masked profound decay and the brutal reality that refused to be ignored—defined the cruel stasis of my addiction. It was a cycle of promise and betrayal, ecstasy, and misery, each feeding the other in a relentless dance. My body became a battleground, my mind a cage, and my spirit a flickering ember resisting smothering darkness. The crack epidemic wasn't just an external plague; it was a parasite inside, a relentless force that bewitched, brutalized, and bound me in chains of my own unraveling.

Looking back, I understand now that addiction's hold was never just about need or weakness—it was propaganda for pain, a seduction of escape, a fight for control in a world that had long since taken it from me. It was a trap laid deep in streets lined with broken homes and broken dreams, a system feeding on despair while pretending to offer salvation. But even at the bottom of that hellish pit, somewhere beyond the tremors and shakes, the paranoia, and the arrests, a tiny spark remained. A part of me refused to be entirely erased by the crack's corrosive embrace. That spark was fragile and flickering, but it was there—waiting for the right moment, the sliver of light that could pull me back from the edge, back from the abyss, toward the possibility of healing and redemption.

Burned Bridges

The slow unraveling of my world began long before the bridges actually caught fire, before the smoke curled thick and black in the sky, suffocating every chance at repair. Addictions crept in quietly at first, dragging me down one careless step at a time, until the walls between me and those who once mattered most started crumbling like brittle chalk beneath relentless storms. Family trust was not something that broke instantly or dramatically in my life; it was a gradual erosion, a betrayed promise every time I showed up late, intoxicated, or not at all, a silent disappointment contoured into weary eyes and tight-lipped warnings. My mother's voice, once warm and steady, had become a weary drone laced with fear and exasperation, her hands trembling when she poured pills into my palm or hurried me to yet another doctor's appointment, hoping to unravel the knot I'd tied in myself without even knowing it. Friends who had once laughed alongside me, shared dreams together in smoke-filled basements or on battered street corners, drifted away in a sorrowful exodus, leaving hollow spaces where camaraderie once lived. Their backs turned not because they wanted to abandon me, but because the person standing in front of them was wearing a new skin—a mask cracked and stained by addiction, erratic outbursts, and elusive promises.

In the beginning, those nights filled with crack and cocaine seemed like the only reprieve from a world that never stopped pressing in. They were my escape pods from blistering reality, but every launch pushed away the people tethered to my orbit. I can still see the glint in my sister's eyes the night she tried to confront me, only for her courage to shatter against my denials and anger. She begged me to come home, to talk, to let someone in, but I had already barricaded myself behind walls woven from shame, rage, paranoia, and dependency. Her voice, once a soothing melody in my chaotic symphony, turned brittle and distant. I remember the afternoon she stopped answering my calls, the knot tightening in my stomach as I realized she had stopped waiting for me to return from the abyss. The phone's silence echoed louder than any argument ever could.

The family gatherings became battlegrounds of unspoken tension. My father, once a gentle but stoic presence, looked at me with that mixture of defeat and sorrow that only a parent feels when watching the very child they once nurtured unravel. His disapproving gaze was a weight heavier than any chain I dragged behind me. My grandmother, whose hands had once held mine so tenderly, declined to meet my eyes, her disappointment wrapping around the room like thick fog. Every interaction carried the sting of judgment, even when words were carefully avoided; every whisper of concern smothered beneath the growing chasm I'd dug between us. I was no longer the boy who once held promise in his smile; I was a ghost haunting their hopes and prayers. The affection that had once knit us together loosened thread by thread until it felt like I was walking a desert, alone and thirsty for a connection that had dried up beyond recovery.

Loss of friendships burned just as fiercely, but they carried a different kind of pain. Friends who'd lived and breathed the same streets, shared the same battles, drifted away not because addiction was a choice I made lightly, but because it turned me into someone unrecognizable. I was that guy who stole money to feed his habit, lied to cover absences, and showed up unpredictably late or not at all. Close friendships dissolved under the weight of broken promises. Conversations that once stretched late into the night about music, dreams, and escape routes became terse and loaded, punctuated by frustration and eventual silence. I remember Mark, my oldest friend, shaking his head after catching me two blocks from the corner, slumped and shaking, fingers trembling from a night chasing highs I no longer controlled. "You're killing yourself, man," he said, but he sounded defeated, like someone who had said the same words too many times to a brick wall. Before long, calls went unreturned, texts left on read, and the only voices that remained were distant echoes from better days.

But it wasn't only friends and family who withdrew—it was the law too, coming down hard like a relentless storm, heightening the isolation. Early arrests brought staredown moments that etched shame deep into my skin. The cold clang of jail cell bars, the scraping of chains, those stark fluorescent lights were brutal reminders that every step off the path had consequences that rippled out like shockwaves, destroying whatever fragile bridges still held on. Court hearings, probation visits, the whispered rumors in the neighborhood about my downfall—all fed a growing narrative that I was a lost cause, someone written off too soon by the very society I'd once believed in.

Convictions didn't just drag my body into confinement; they dragged my spirit farther away from the people who had stood by me. I can't forget the nights in holding cells where booths of fluorescent light mocked me like the spotlight of shame. It was in those cold hours of reflection, stripped down by cold walls, that the weight of what I'd lost truly pressed down, crushing the last flickers of hope. The faces of my mother's tears, the blank looks from childhood friends, the missed phone calls formed into a never-ending loop of regret in my mind. I wanted to claw my way back, but the weight of addiction was like quicksand—it pulled me under at every desperate grasp.

Even among other lost souls behind bars, the sting of isolation was sharpened by the invisible scar of ruined trust. Conversations were tentative, friendships fragile, and when those moments of connection bloomed, they evaporated too fast, consumed by the shadows lurking beyond every corner. I didn't belong in those faces either, filtered through the lens of my downward spiral and faltering loyalty. I felt like a ghost walking through a nightmare painted with my own mistakes, the haunting refrain of "what if" echoing louder than the clanging metal doors.

Back home, every missed parole appointment, every slang-laced lie to keep the peace, every relapse stole another piece of faith from those who cared. Family calls became less frequent until they faded to silence; the phone that used to ring with love became a cold instrument that never sang my name. Cousins stopped inviting me to family barbecues; my aunts avoided me in the store; even the neighborhood kids looked at me with wary distance. I turned into someone who sunk lower and lower, dragging my family's name through the mud despite their desperate hopes to see me rise. The realization that my actions didn't just hurt me was slow but shattering—a toll paid in the currency of broken hearts and fractured lives.

One night, after a run-in with the cops turned ugly, I found myself slumped on a cracked stoop, the cold concrete pressing into my skin as rain began to fall like icy needles. The sky opened up, and the water sluiced over me, but even the cleansing rain couldn't wash away the grime of betrayal that clung to my soul. Flickers of memories surged—my sister's voice calling out in anger, my mother's tears, the vacant expressions of former friends. I understood then in a raw, undeniable way how addiction didn't just consume the user; it spiraled outward, tearing through the fabric of relationships, chewing up trust and spewing out smoke and ash in its wake.

The bridges I'd burned weren't just made of words or moments—they were constructed from shattered hopes, stolen time, broken promises, and the ruthless demands of a demon that whisperingly demanded everything. Every attempt to reach out, every apology whispered amidst fuzzed-out moments, met walls of suspicion hardened by years of disappointment. Family gatherings became battlegrounds where my presence was either tolerated or feared, a constant reminder of battles lost and scars that refused to heal. I was treated less like a son or a brother and more like a stranger, a cautionary tale whispered about in hushed tones—a ghost whose existence was both a burden and a sorrow.

This loss of trust went beyond mere distance; it was a complete severance, an unraveling of the invisible threads that had once tied me to meaning and belonging. It felt like drowning in a sea where no hands reached back, where each cry for help was met with silence or suspicion. The isolation fed the addiction even further, creating a vicious cycle where loss fueled despair and despair fueled more destruction. I was trapped in a spiraling nightmare, watching connections crumble, friendships dissolve, and family ties fray, powerless or unwilling to stop the collapse.

Even as addiction sharpened its claws deeper into my flesh, I remember moments of flickering clarity and the gut-wrenching sadness that came with seeing the ruins of what I'd once held dear. There was no dramatic climax that severed the ties, no explosive confrontation, but rather a series of small, grinding betrayals—a missed birthday, a lie told to cover absence, a phone call hung up mid-sentence—that eventually piled so high they could not be climbed. And once those bridges were gone, once the fires consumed the linkage, the cold darkness felt ready to swallow me whole.

The descent into isolation was terrifying in its quiet relentlessness; there were no cries for help loud enough to break the silence, only the gradual void that stretched between me and everything I had once loved. The very people who had been my anchor, my safety net, faded into the background like ghosts at the edge of vision, their images blurred by the haze of smoke, drugs, and self-destruction. My world became smaller, tighter, darker—sometimes no more than a needle, a pipe, a street corner where the night swallowed me whole in its cruel embrace.

Yet in the absolute wreckage of burned bridges and scorched relationships, an ember of something fragile still flickered beneath layers of ash. The piercing loneliness and the weight of loss sometimes shook me awake at night, forcing a painful reckoning with what I had done and what remained. The shadows of my past love, once so vibrant, seemed to reach through the darkness, reminding me that beyond the ruin lived a possibility—one that would demand that broken trust be rebuilt brick by agonizing brick, that redemption was not a magic spell but a painstaking act of persistence and courage.

Burned bridges leave scars on the landscape of the heart. They are reminders of the fragile nature of human connection and the devastating effects addiction can wield in its relentless course. They paint the portrait of a man lost in chaos, buried underneath the cigarette smoke and dissolved promises, a man still yearning to come home—to reclaim the love, the trust, and the family he had once unwittingly set aflame. The truth I came to live with was harsh and unforgiving: losing the people who believed in me the most was the heaviest burden of all, and the first battle I had to face if I ever wanted to step back into the light.

Run-Ins with the Law

The night always had its own rhythm, pulsing with danger and desperation, like the twisted beat of some broken, underground drum that refused to stop. Streets alive with whispers of raids, the scuff of hurried footsteps on cracked pavement, and the sharp click of handcuffs cutting into wrist flesh—those were the sounds that followed me like a shadow that wouldn't leave. By then, my life had become a tightrope stretched thin over a bottomless pit, where every wrong step plunged me deeper into chaos. The phrase "run-ins with the law" barely captured the raw, suffocating weight of each arrest, each violent brush, each moment when the world seemed ready to shatter beneath the force of a handcuffed fate.

Back when I first slipped into heavy drug use, it wasn't like I woke up one morning and said, "Today's the day I lose everything." Nah, it was a jagged descent, a blurry slide down a slope slick with mistakes nobody warned me about. The first time the cops slammed me against a dirty brick wall, the scent of stale urine and fear thick in the air, I fooled myself into thinking I was just unlucky, caught in someone else's crosshairs. But luck wasn't the game anymore. It was survival — brutal, unrelenting, and always on the edge. I still remember the snap of the handcuffs, biting into my wrists as the officer's voice barked commands just loud enough to echo in my skull but far too muted to drown out the pounding in my chest. That was the sound of my own heartbeat announcing my surrender.

It wasn't just the law; it was the violence that stalked every corner, every alleyway, every stolen moment under the cracked streetlights. I wasn't just running from cops—I was running from the ghosts of deals gone wrong, from vendettas fueled by desperation and rage, and from a past that refused to let go. One night, after a botched drug exchange, I found myself backing into the cold steel of a shotgun, the barrel pressing against my temple, the man's eyes wild with suspicion. My breath hitched, warm against the freezing metal, the world narrowing to the tremor of his finger tightening on the trigger. Somehow, a mix of luck and numbness saved me that night, but the brush with death clung to me like an unwanted tattoo, a reminder that every choice, every step, could be the last.

Those days morphed into a collage of blurred memories: police sirens slicing through the predawn haze, fluorescent lights buzzing in holding cells, the endless parade of faces all twisted in worry or indifference. Some of them were friends, caught up in the same mess, their eyes hollowed by the same fight to just catch a break, or catch a breath. Others were strangers, acquaintances from the streets or the corners where we hustled, always ready with a quick excuse or a bitter joke to mask the fear that chased us down every alley. Inside those jail cells, amid the stench of sweat and crushed dreams, time fractured. Hours stretched endlessly, each second ticking off a countdown not toward freedom, but toward the inevitability of facing those same streets again, scarred and broken.

I remember the sting of the cuffs in the rain, rain that washed the filth off the gravel but couldn't touch the filth in my soul. Arrests piled up like unwanted souvenirs, badges of a war I never wanted but held no power to stop. Each encounter with the law chipped away at whatever remnants of innocence remained, seeping into conversations with my family, where their voices cracked between worry and anger. I wasn't the kid they knew anymore—not after the nights spent behind bars, the days withered in courtrooms, the endless cycle of release and re-arrest that felt less like a punishment and more like a sentence to a life unchosen. The strain twisted relationships until the words were brittle, and the silence that followed was louder than any scream.

Violence wasn't just an outside force; it was a language I learned to speak fluently, like a native tongue. The crack epidemic taught us brutality with a harsh dialect—gunshots replaced lullabies, and broken bodies replaced hugs. I witnessed fights that bled into the streets as paint spilled on concrete, friends turning enemies over territory, or loyalty shattered by betrayal. Once, I was grazed by a bullet in a skirmish over a two-dollar debt that seemed too small to ignite so much fire, but did anyway. The pain shot through my side sharp and raw, but the fear deeper, because I knew the bullet was not just trying to kill flesh—it was trying to kill hope. It was aching, a reminder that survival wasn't guaranteed, only earned minute by minute, breath by breath.

Every time the blue and red lights swept over my face, there was a split second when the future condensed into a single, frozen frame. Would I be taken in? Would there be a fight, a struggle? Would this arrest be the last, or the first step on a darker path? Freedom became a form of currency I spent recklessly, always running a debt I knew I couldn't repay. The courts became a routine theater: judges with eyes glazed by boredom, lawyers with their detached kindness, prosecutors with their hungry glares. I learned to speak in legal jargon I barely understood, nodding through the pleas and sentences that felt like they were directed less at justice and more at punishment designed to break. Jail wasn't a place for healing, it was a furnace that tested every ounce of resilience, where violence still lurked behind steel bars and beyond the watchful eyes.

One memorable encounter saw me thrown into a holding cell with a man whose fury didn't come from addiction or survival, but from a darker corner of human torment. His rage was a physical force, a storm in human form, and in the desperate confines of that cell, he unleashed it with a violence I can still hear when I close my eyes—a howl scattering against the concrete, a rage without end or reason. It shook me, fractured what little sanity I had left, and planted a seed of fear that night would never be just night again. I learned to mask the tremors in my hands, the scream caught in my throat, to survive not just the law's cold grip but the savage heat of other men's pain.

The loss of relationships during this period etched scars deeper than any bullet hole. Friends became more distant, avoiding calls or flicking away texts like burning coals. The few who stuck around carried their own burdens, making companionship fragile and transient. My family, once a source of light, now seemed like a distant shore I was drifting away from with each arrest. Every visit home became a test of will, of masks worn to hide the truth, the hustle to keep from shattering their fragile hope. The holes my addiction gnawed through trust made reconciliation seem more like a fantasy than a possibility. I felt like a ghost haunting their memories, a reminder of promises made and broken.

With every court date, every stint in jail, the precariousness of survival tightened its grip. The streets weaved a tangled web of predators and prey, and sometimes there was no clear line between the two. Running from law enforcement didn't mean escaping danger; it meant diving deeper into a world where violence was currency and betrayal was business. The latter that I knew too well as friends turned to rats, deals fell apart with a gunshot's warning, and every opportunity carried the risk of death or arrest. I had friends vanish overnight, swallowed by the system or the streets, their names whispered in mournful tones. Each loss was another piece of the fragile mosaic of my life crumbling away.

Those run-ins weren't just about dodging arrests or surviving violence—they were crucibles refining my understanding of the world and my place in it. I saw the faces behind the badges—men and women burdened with their own fears and limitations, enforcers in a system built on punishment more than redemption. I glimpsed the cyclical nature of pain, how neighborhoods strangled by poverty and drugs bred despair that fed back into violence and incarceration. It wasn't black and white; it was a toxic grayscale that dulled every edge to a numb haze. My own reflection in the cracked storefront windows was a stranger I barely recognized, a man caught in the web of his own making yet desperate to break free.

In quieter moments, between arrests and scars, I thought about what survival really meant. Was it just a measure of how long I could evade the law, how many bullet holes I could dodge? Or was it something deeper—a tether to hope, however faint, pulling me out of the spiral? That hope was fragile, obscured by the haze of drugs and violence, but it flickered still. When the law knocked on my door, pulling me from the streets I called home and condemning me to cold cells or crowded courtrooms, it was also a chance—however small— to choose a different path. But those chances were rare, and the price was steep.

Every run-in brought me closer to a breaking point, a place where system and street collided in a brutal dance that left no room for mercy. I became an expert in survival, not because I wanted to be, but because I had no choice. The sting of arrests became routine pain, the numbness blurring lines between freedom and captivity. Yet, in that grinding chaos, somewhere underneath the grime and the fists that knocked me down, there lingered a seed of something that couldn't be crushed—a will to live, even if I didn't yet know how to live well.

I'd been arrested so many times by then, lost track of the count, each one a shallow dent in the hull of a ship already leaking fast. Some nights, the jail cells felt like a refuge, a place to escape the relentless demands of the streets if only for a little while. But every time freedom came calling, I had to face the same cruel truth: the streets wouldn't forgive, wouldn't forget, and the law was always watching, ready to snatch me up again. It was a crucible of desperation and determination, brutality and fleeting mercy, a battleground where my life unfolded with all its jagged edges exposed.

I lived in a world where survival was a game with no practice rounds, where every encounter with law enforcement was a test not just of luck but of who I was underneath the scars and addiction. Those run-ins shaped me, forged a hardness that could break but also protect. They exposed the precariousness of existence on these streets where the crack epidemic had already bled out too many souls, where every arrest was a chapter in a story no one wanted to read, but everyone in the neighborhood knew by heart. In that dark urban symphony, my own story played out between sirens and silence, between cuffs and fleeting freedom—always on the razor's edge, always risking the fall.

The Darkest Night

The night stretched on like a thick, suffocating fog, swallowing every sliver of light and hope. It was the kind of darkness that didn't just cloak the streets but seeped deep into the bones, marking all those who dared to walk its dangerous edge. The air hung heavy with tension, a quiet prelude to chaos. I had been sinking deeper into the abyss—more into the numb embrace of crack than into the life I once dreamed of, a life left shattered like the fragments of a broken mirror reflecting my fragmented soul. Every familiar face around me had faded like ghosts in a dying fire, some lost to jail cells, some to streets soaked in violence, others to graves prematurely dug by the unforgiving hands of addiction. My world had contracted into a series of swift trades, harsh words, and fleeting moments of a twisted high. There were no tomorrows anymore, only urgent, desperate nows.

That night lingered with a foreboding that I almost ignored, but deep down, a part of me felt the ominous shadow looming closer. I remember sitting against the cold brick of an alleyway, the roughness biting into my back, the sting grounding me even as my head swam with the toxic euphoria of another hit—the crackpipe pressed greedily to my lips, drawing in the poison disguised as escape. Every inhale was a steady surrender, a small death, until my chest felt hollow and my thoughts looped endlessly in fractured, dark spirals. Around me, the city was a cacophony of sirens, distant shouts, and the occasional thud of someone else's pain echoing in the night. The rhythm of a hard, unforgiving world played loudly against the flickering streetlights that illuminated the cracked pavement and the fading graffiti—barely visible relics of better days. But my focus was blurred, captive to the temporary solace that the drugs promised but never truly delivered.

In that haze, I hardly registered the footsteps before they came— a deliberate, predatory pace that sliced through the murmur of night. I felt it first as a shadow falling over me, and then a cold, metallic weight pressed against my temple. The world seemed to shatter in an instant, exploding in a rush of adrenaline and horror. The barrel of the gun was heavy, unyielding, and certain, pressing into my skin with a promise of finality. Time slowed down, each moment stretching into an eternity where fear, pain, and disbelief collided in a chaotic whirlwind. I remember the harsh whisper of broken words, threats spat like venom, mixing with the roar of my pounding heart, every beat a countdown to either death or escape.

The crack epidemic had turned my neighborhood into a battleground where respect was measured in violence and survival was the currency for each breath. I was no stranger to conflict, no stranger to danger, but that night the stakes felt infinitely higher, the line between life and oblivion blurred by a trigger finger's twitch and a heart hardened by too many scars. What led to that moment was a tangled mess of grudges—alliances broken, debts unpaid, and betrayals sewn deep into the fabric of our reckless existence. I didn't even recognize the man who held me hostage to his fury; his eyes were wild, filled with the same desperation that had consumed me for so long. It was as if we were both drowning in pain, but only one of us was willing to watch the other sink.

The gunshot tore through the stillness, a cruel punctuation that left my body trembling on the edge of destruction. The impact wasn't immediate; it was a slow, searing burn as the bullet found its way, ripping through flesh and shattering bone, a brutal punctuation to a life too long teetering on the edge. Pain exploded in waves, wrenching me out of the drugged stupor and forcing raw awareness down every nerve. Blood blossomed warm and sticky, a vivid reminder that I was still alive, still breathing despite the odds stacked so heavily against me. The alley became a spinning vortex of sounds—the frantic slam of footsteps retreating, distant sirens growing closer, the harsh commands of voices rushing to intervene. My world had narrowed to a tunnel of agony and shock, as if the night itself was holding its breath, waiting to see if I would slip away.

In those agonizing moments sprawled beneath the indifferent city sky, I confronted the darkest corners of myself—the part that believed life was nothing but a series of bad choices, a downward spiral with no way out. But somewhere, beneath the pain and the numbness, a flicker of something ancient yet fragile stirred—a whisper of the will to survive. The paramedics arrived, their faces grim yet steady, voices calm as they worked quickly, binding wounds and pumping life back into my failing system. The cold glass of the ambulance window separating me from the street life I had known was like a barrier between death and whatever remained of hope. Every siren scream felt like a call from the abyss, pulling me toward judgment, but also reminding me of a chance—a shaky, uncertain chance to rewrite my story.

The hospital was a blur of white walls, sterile lights, and the constant hum of machines that charted the fragile dance between life and darkness beneath my skin. I was trapped in a haze of painkillers and fever dreams, memories slipping in and out like fragmented scenes from some twisted movie. Nurses moved around me in practiced rhythms, voices soft but urgent. Strangers touched my skin with professional care, the warmth of human compassion cutting through the cold shroud of despair that had settled so deeply in my heart. As I lay there, the weight of that near-death moment pressed on me like a suffocating shroud—an impossible silence filled only by the harsh whisper of my own ragged breathing.

Family and old friends came and went in muted shadows. Their eyes reflected a mix of sorrow, disbelief, and fragile hope, but it all felt distant, as though I was floating above myself, watching a ghost replay scenes of a life fractured beyond repair. The violence and addiction that had once seemed like insurmountable forces suddenly took on a human face—the faces of those who refused to give up on me, who saw not a lost cause but a flickering ember waiting to be fanned back into flame. Yet with every visitor, I felt the echo of all I had lost—the trust broken, the bridges burnt, the love abandoned in the wake of my self-destruction. It was a nightmarish reckoning, a brutal inventory of pain and regret.

Amidst the sterile walls and endless hours of waiting, a profound clarity began to take shape within the haze—a raw, unsettling understanding that this was a reckoning far beyond any street fight or drugside haze. It was a crossroads, a moment forged in extremes where death had come knocking not just on the door of flesh but on the soul itself. The near-murder was not just a physical attack but a devastating mirror reflecting the consequences of a life lived without reins. I realized that the chain of pain I had been caught in stretched far beyond myself—it wrapped around the neighborhood, the families, the silent victims lost in a world where violence was currency and addiction was escape.

In that darkest night, ironically, the seeds of transformation were planted. The brush with death peeled back layers I had long hidden beneath bravado and denial, exposing raw, aching truths I could no longer ignore. The image of my own blood seeping into the cracked concrete stayed with me—an undeniable testament to how close I had come to vanishing completely. What haunted me most wasn't the physical pain, but the haunting question: What happens to a man when he's left with nothing but the ruins of his past and a fragile chance at survival? Could he rise from the ashes, or was he forever condemned to the shadows?

The recovery was agonizingly slow. Pain pills became a new kind of drug, tempting in their siren call, threatening to pull me back into the same abyss. Every waking moment was a battle between despair and the thin thread of resolve, stubbornly holding on. Police interrogations wove through the days, some suspicious eyes searching for answers, others more interested in filing reports than understanding what made a young man spiral so far down. The jailhouse was never far from my mind, a looming specter hovering over every step toward healing. I was forced to confront not only my physical vulnerabilities but the emotional shrapnel that pierced deeper than any bullet.

Interwoven with the daily grind of therapy sessions, medical treatments, and legal entanglements, the memories of that night—sharp, vivid, relentless—demanded I face the full weight of my choices. Old allies turned away, some whispered rumors of vendettas unfinished, and I knew that surviving the shootout was merely the first battle against a lifetime of ghosts. The narrow escape from death underscored the fragility of my existence and the dangerous paths I had embraced, paths lined with broken promises, shattered loyalties, and the lure of easy money that always came at an unbearable cost.

Yet, amidst the storm, there was a growing current of something unfamiliar: a yearning not just to survive but to reclaim meaning. To not become another statistic swallowed by the streets or the legal system. That night's brutality carved a deep scar, but it was also a grim gift—a brutal realignment of priorities. Like the echo of a long-forgotten drumbeat, the memory of blood and fear pulsed relentlessly, urging me toward change. The world I had known—the gangs, the drugs, the endless nights chasing oblivion—looked suddenly hollow, a brittle house of cards waiting to collapse.

As days turned into weeks, the physical wounds slowly healed, but the battle raged fiercest within. Nightmares haunted my sleep, replaying the gun's cold shadow, the whispered threats, the cold weight of imminent death pressed to my skin. Yet with every terrifying echo, resilience hardened its grip. The presence of a therapist—someone who saw me beyond the scars, beyond the street tale—became a lifeline, a mirror where I could begin to piece together the fragments of myself. Through sobs and silence, I started to voice the pain that ran deeper than the surface wounds, opening a delicate door to forgiveness and understanding, first for others, then painfully for myself.

The darkest night marked not just survival, but a piercing awakening. It cracked open the walls I'd built around my heart and mind, confronting me with the raw truth that life, fragile and uncertain, was still mine to own. From that threshold, tentative steps toward recovery began—not in grand gestures, but in the small, grueling moments of choice: to breathe, to fight, to hope. It was a turning point forged in blood, shadow, and fire, a brutal reminder that sometimes the only way out of darkness is through it. And as the dawn finally broke, pale and unassuming, I found myself standing at the fragile edge of a new beginning, a chance to transform pain into purpose.

Chapter 4
Prison Walls

Locked In

The clang of the heavy metal door slammed shut behind me, sending a cold ripple down my spine. The sound echoed through the steel and concrete corridors like a death knell, sealing me off from the world I once knew. That first moment inside the compound was suffocating; a claustrophobic weight pressed deep into my chest, a tangible reminder that freedom had become an illusion. The air hung thick with the acrid smell of sweat, bleach, and something metallic, a unique perfume forged by the hundreds of men locked away in this place. It was a harsh reminder that this new world operated on unforgiving terms, where every second demanded vigilance and every glance could carry silent threats. As I was ushered into the cell, the reality of prison life settled over me like a dense fog — unfamiliar yet ominously familiar, like stepping into a grave I dug for myself long ago.

The walls were bare and unyielding, a harsh contrast to the life I left behind, with its chaotic mixtures of hope and destruction. Here, the only color was the dull gray of concrete and rust. Even the small window, barred and smudged, seemed to offer no promise but rather to mock the idea of daylight and freedom. I sat on the thin bunk, a slab of cold metal covered by a thin mattress that did little to cushion the bones beneath. The silence was deafening, broken only by the distant murmurs of other inmates and the clinking chains of far-off guards. In those first hours, I was left alone with my swirling thoughts, a whirlwind of regret, anger, and a numbing sense of disbelief. How had my life twisted so sharply, spiraled into this boxed-in existence where every choice would be scrutinized, and every moment counted against me?

The transition was brutal. I was not prepared for the stark lack of privacy, forced proximity with men whose faces told bloodied histories of their own. These were warriors of the streets, some hardened by years behind bars, others just as lost as I was, clinging desperately to any semblance of control in a world where power was often measured in violence and fear. The hierarchy was palpable. It didn't matter who you were on the outside; here, you had to earn your place or suffer the consequences. I learned quickly that a wrong word or a misread signal could ignite a wildfire. The early days were a delicate dance—a balancing act between asserting a fragile sense of self and blending into the background enough to avoid unnecessary attention. I felt like prey, exposed and vulnerable in a jungle where survival depended on understanding unspoken rules and codes.

Every interaction was a negotiation of power. Prison, I realized, was more than physical confinement; it was a mental battlefield. In those tight spaces, with the presence of constant surveillance and few outlets for expression, the smallest gestures carried enormous weight. I observed the way men measured each other's strength not only through fists but through silence, stares, and stubborn pride. My mind raced to adapt, to find strategies that could keep me safe yet not so aloof as to draw suspicion. I became a watchful eye and a listening ear, deciphering the intricate web of alliances, rivalries, and grudges that defined the facility's ecosystem. Time bent strangely there, the days blurring together as routine and monotony meshed with constant undercurrents of tension. I learned to count the passing hours by the staggered arrival of meals, the clang of the metal gates opening and closing, a slow cadence marking my days served in a place that felt designed to strip the soul bare.

With every passing moment, memories of past choices flooded in with renewed intensity. I was haunted constantly by the echoes of my younger self—the boy who chased quick survival in back alleys, who had believed that toughness and bravado could shield him from the fractures growing inside. The faces of those I had lost—friends wounded by violence, family strained by my absence—weighed like chains on my conscience. The streets whispered their siren song in quiet hours, tempting me with false promises of belonging and escape, even though I knew the depths of destruction they held. I replayed decisions like rerun scenes in my mind: moments when I'd crossed lines, when pride blinded me, when desperation pushed me into recklessness. These regrets became an internal prison of their own, as confining as the concrete walls surrounding me. I was forced to confront the raw truth that every tough choice made to survive had also built the cage I was now trapped in.

Yet, amid the relentless harshness of confinement, moments of clarity sliced through the haze. The stillness allowed for introspection I had never before dared to entertain. Without the distractions of the streets—the noise, the adrenaline, the constant scramble for status—I began to understand the depth of my wounds. Addiction had been a siren leading me toward destruction, offering temporary relief but insidiously tightening its grip until I was nearly lost in its depths. Prison offered an odd kind of refuge: a space where the chaos of my external world was temporarily muted, forcing me to reckon with the turmoil inside. This confrontation was terrifying but vital. I felt the first flickers of an uneasy honesty—a fragile seed planted in barren ground—that perhaps I could be more than the sum of my mistakes. But such realizations were bitter medicine, twisted with frustration at how far I'd fallen and how much work lay ahead if I ever hoped to reclaim my life.

The physical toll of the environment was brutal, shaping every day into a battle of endurance. The body ached constantly—tight muscles clenched from tension, joints stiffened by cold nights on unyielding cots, skin raw from the grime and harsh cleaning agents that never seemed to wash away the invisible filth. The food was meager and industrial, devoid of comfort and nourishment, a reminder that the facility was designed not for healing but for containment. Even the limited exercise yards were arenas of unspoken contests, places where fights could ignite in seconds over the smallest perceived slights. Sleep was an elusive prize, shattered by echoes and footsteps in the hallways, the low murmur of distant conversations, and the gnawing worries that no cocoon of darkness could silence.

Over time, I developed coping mechanisms—rough shields against the assault of prison life. There were casual conversations exchanged with men behind bars who, despite everything, offered flashes of kindness and understanding. Shared stories became a strange form of connection, reminding me that beneath the hardened surfaces were fractured souls seeking redemption, recognition, and sometimes simply to be seen. I grew to value these moments, tiny islands of humanity in a sea of despair. At the same time, I steeled myself to face the ever-present threat of violence. Learning how to defend not just my body but my dignity became a daily lesson. I realized that resilience was not just about strength but also adaptability—the ability to navigate complexity without losing sight of the core self.

Mental health threats loomed large. Days without hope chipped away at morale; depression lurked in the corners of my mind, and occasional bouts of panic clawed at my chest when the weight of the past collided with the oppressive walls around me. There were nights I lay awake, wrestling with demons more cruel than any inmate. How do you forgive yourself when the world has echoed with your failures? How do you muster the will to change when every instinct is to survive a colder, harsher reality by any means necessary? It was in those moments of isolation that the true battle of prison unveiled itself—not one fought with fists, but one waged within the shadows of the soul.

Despite all this, I found myself reflecting on the twisted ironies of imprisonment. The streets had promised freedom through rebellion, yet delivered bondage through addiction and violence; here behind bars, my body was physically confined, but my mind was no longer on the run. I could see the contours of my life with brutal clarity for the first time. The recklessness that had once masqueraded as strength now appeared as youthful blindness; the pride that had fortified me was a cage of ego, cutting off the possibility of growth. This painful self-examination laid the foundation for a gradual transformation. Each day offered a choice—to sink further into despair or to reach for something beyond the suffocating concrete walls.

The system itself was a reflection of broader societal failures, a reminder that my imprisonment was not just a personal downfall but also a symptom of larger broken structures. The overcrowded cells, the lack of adequate mental health support, the revolving door of release and re-arrest—these were parts of a cycle that ensnared so many men like me. It was easy to feel invisible, lost among statistics and labels that ignored the human beings behind the numbers. Yet, even amid this despair, I understood the importance of this moment—not just for survival, but as a crucible where I could begin to forge the first glimmers of redemption. To do so, I would need more than grit; I would need to learn trust, seek help, and confront the pain I had so long run from.

I began paying close attention during the sparse times for education and counseling programs offered in the facility. These moments felt like lifelines, opportunities to glimpse a different path. I listened to the stories of men who had walked similar roads, whose survival meant more than mere endurance but also learning and change. The therapists and volunteers challenged me with hard questions that forced me to peel back layers of denial. Slowly, the walls around my mind began to crack. I realized that survival here required more than brute strength and fear; it required confronting vulnerability, admitting faults, and daring to imagine a life beyond the bars.

There was an almost cruel rhythm to the days, where the monotony gave way to unpredictable bursts of chaos—an outbreak of violence in the yard, a sudden lockdown, a whispered warning that could mean the difference between safety and harm. In navigating this volatile environment, I developed a sharper sense of self-awareness. I learned to recognize the patterns of behavior that led others to fall into traps of aggression, and to walk the fine line between caution and courage. I trained myself to manage anger and frustration, understanding that in this world, explosive outbursts could escalate into catastrophe. This self-control was essential, a new kind of power that began to replace the reckless bravado born in the streets.

Relationships inside were complex; trust was fragile, and alliances could shift like sand. Some men became reluctant allies, providing a buffer against the most dangerous elements, while others remained distant, locked behind their own scars and fears. There were moments of fleeting friendship that shimmered bright but often threatened to dissolve in the brutal reality around us. Even those with the hardest exteriors sometimes cracked, revealing glimpses of pain and humanity. I learned to listen beyond the surface, recognizing that everyone was carrying a burden, albeit one shaped by different experiences. This understanding fostered patience and empathy within me, and slowly chipped away at the bitterness I had held onto since my youth.

My reflections often turned to family, the few connections left strained almost to breaking. Letters from home were lifelines but also sources of anguish—reminders of the damage my absence inflicted and the distance that seemed impossible to bridge. I wrestled with guilt, aching to be a better son, brother, or cousin, but aware that time and choices had erected walls just as formidable as the prison cells. The thought of reconciliation felt both distant and desperately needed, a light flickering faintly in a dark tunnel. I promised myself that if given a second chance, I would build those bridges anew, but first I had to confront the man prison was molding me into.

Despite everything, there were moments that sparked hope. A piece of poetry scribbled in a notebook, a shared laugh in a rare moment of levity, the silent endurance of men who refused to be crushed by circumstance. These slivers of light hinted at the possibility of change, not just for me, but for everyone caught in this cycle. I began to see that prison, though brutal and dehumanizing, could also be a crucible for transformation if one was willing to face the darkness and seek out the fragile threads of hope within the shadows. This understanding became a lifeline in itself, a reminder that even in confinement, the human spirit could flicker with resilience.

Yet the fear of falling back into old patterns was always lurking. I knew that the world outside remained unforgiving, that without preparation, freedom could feel as suffocating as imprisonment. The scars of addiction and violence ran deep, and I was painfully aware that my journey toward healing was just beginning. The prison walls were not just physical barriers but metaphors for the internal prisons of shame, addiction, and anger. Breaking free from these would take more than time; it would demand courage, honesty, and perhaps above all, forgiveness—of others and of myself.

The days stretched on, each a test of endurance and will. Slowly, I started crafting a new identity within these concrete confines, one built on small victories and hard truths rather than bravado and denial. I picked up books to read, attended group sessions when possible, and wrote letters that spilled out the tangled emotions I struggled to voice in conversation. These acts, small but significant, became steps toward reclaiming a sense of agency in a place designed to strip it away. I recognized that this process was messy, fraught with setbacks and moments of despair, but also filled with the possibility of emerging—if not whole, then at least enough—on the other side.

In this crucible of incarceration, I began piecing together the fractured parts of my life. Each day behind bars was both a punishment and a chance, forcing me to confront uncomfortable truths and shaping the fragile contours of a new beginning. The prison was a harsh mirror, reflecting both the shadows of my past and the glimmer of a future yet unwritten. Locked in, I was also locked into the hard work of transformation, a journey that demanded every ounce of courage and honesty I could muster. And though the walls remained cold and unforgiving, within them grew a quiet, determined resolve: to one day step out not as the boy who fled into streets of chaos, but as a man ready to reclaim his life and rewrite his story.

Isolation and Reflection

The iron bars cast long shadows across the narrow cell, the faint fluorescent light flickering intermittently as if uncertain whether to stay alive or die. The air was heavy, stale, and cold, each breath a reminder of the suffocating constraint of these walls. In this small, confining space that barely held his body, the real prison was not his physical surroundings but the relentless storm raging beneath his skin—the barrage of memories, regrets, and fears that kept him tethered to a place far darker than the one his eyes could see. Days blended into nights without distinction. The enforced solitude, while brutal, birthed a silence so deep it screamed louder than any siren on the streets. It was in this silence that his mind became both jailer and confessor. Forced still by the rhythms of prison life, he found himself pulled into a crucible of self-examination, wrestling with the shadows of his past as they loomed larger with each hour stretched thin by confinement.

The clang of the steel door shutting behind him echoed like thunder in his chest—the cold mechanical finality translating into a heartbeat that drummed the litany of mistakes made. The initial disorientation of being thrust into this alien world gave way to a crushing sensation of isolation, not merely physical but existential. He was separated from everything he once knew—family, friends, the chaotic streets that had been both friend and foe—left alone not only with the absence of others but with the undeniable company of his own thoughts. The walls became mirrors, reflections twisting in the dim light to confront him with the man he had been and the man he feared he might never escape becoming. The faces of those he betrayed and those he lost floated unbidden, each memory cutting with the sharpness of glass. He traced back the endless chain of decisions: the first hesitant step into gang life, the allure of brotherhood forged in the fires of street violence, the inescapable drag of addiction pulling him deeper into an abyss where hope was a currency long devalued.

Each night, the cell seemed to close in tighter, the cold biting into his skin, but the chill inside his mind far colder and more relentless. With no distractions other than the ceaseless passage of time marked by distant, muffled sounds of prison routine, his mind became a battlefield. He had thought he understood pain—physical, emotional, survival—but here it was distilled to pure essence. Pain without end, pain that stripped away pretense until all that remained was raw, aching vulnerability. His self-contempt clashed fiercely against a fragile, stubborn flicker of something unbroken deep inside. It whispered not in grand promises but in subtle ways: a stubborn refusal to surrender fully to despair, a tiny ember of self that clung to the possibility of renewal. Yet the path to that renewal was obscured by thick fogs of shame and regret, each recollection a stone in a backpack dragging him deeper into a stagnant pool of guilt.

The slow, grinding monotony of prison life forced a kind of brutal honesty. There were no places left to hide behind—the bravado, the defiance, the carefully cultivated mask of invincibility all stripped away under the unyielding lens of isolation. He confronted truths about himself that had been easy to avoid when surrounded by chaos and noise—the depths of his addiction, the real human costs of his violence, the faces of betrayal etching guilt into his soul. Each confrontation was a harsh lesson in accountability, a grueling process of naming each wound and scar with brutal clarity. For the first time, he began to peel back the layers of denial, recognizing how much pain he had inflicted on others and himself, how deep the fractures ran beneath his calloused exterior. The rage that had bubbled inside him no longer had a convenient outlet; in the quiet, it morphed into a burning shame that stoked the fires of a desperate desire to change.

The bitterness of confinement was amplified by a profound loneliness. The few interactions he had—guarded conversations in the yard, fleeting exchanges during labor assignments—felt like fragile threads barely holding him to the world beyond these walls. Yet even these interactions were pregnant with unspoken fears and cautious suspicions, the logic of survival in such a place demanding masks just as carefully worn as those outside. It was a loneliness marked not just by physical separation but by a deeper estrangement from his own sense of self. The boy who had wandered into gang life with uncertain steps was not the man sitting cross-legged on the cold concrete floor, yet neither was he the redemption he dreamed of becoming. He dwelled in a liminal space, caught between what was and what might be, haunted by the past's relentless echoes yet tentatively grasping at the fragile shoots of hope sprouting in moments of clarity.

In the midst of this crucible, he began to explore the architecture of his pain, tracing the roots of his addiction and violence back to broken promises and deeper wounds unspoken. Memories his mind once pushed to the margins surfaced with new, harsher light—the absence of father figures, the pervasive trauma of poverty, the trap of systemic neglect that had funneled him into deadly choices. He saw his story no longer as isolated failures but as fragments of a larger, heartbreaking mosaic shaped by forces far beyond one man's will. This realization did not excuse the damage done, but it offered a doorway to understanding—a crucial step toward self-forgiveness and growth. Yet the path remained steep and littered with obstacles; old habits and impulses clawed at his resolve, the demons of addiction whispering seduction even in chains. Nightmares chased sleep, and waking hours were shadows thick with doubt. Every small victory was a battle he had to fight twice: once against the external confines, again against the internal collapse.

Still, it was within this prolonged silence that a kind of fragile strength began to root. The very emptiness that threatened to consume him also allowed space for reflection, allowing him to hear his own voice beneath the clamor of street life and rushing blood. The pages of his mind became a canvas for painful self-inquiry and cautious hope. He returned again and again to the question that would haunt and drive him: Who was he beneath the layers of violence, fear, and addiction? What, if anything, lay beyond the endless cycle of destruction that had claimed so much? These questions did not come with immediate answers but with the slow dawning of self-awareness—an understanding that change was neither instant nor guaranteed, but a process made up of countless small, sometimes imperceptible steps. He learned to sit with discomfort, to witness his own failures without surrendering to them, to hold pain not as a punishment but as a teacher.

Amid his isolation, he began to reach inward and outward simultaneously. Journaling became a lifeline, the scrape of pen on paper a small rebellion against the erasure of identity. Writing allowed him to translate the chaos in his mind into an ordered sequence of thoughts and feelings, providing a refuge where honesty was possible and where the beginnings of a new narrative could take shape. The act of telling his story, even if only to himself on these pages, chipped away at the thick armor of silence that had long kept him imprisoned within his own thoughts. He visualized—sometimes with fierce desperation—the possibility of life beyond these walls, beyond addiction, beyond the violence that had marked his past. He imagined forgiveness, not as an abstract concept, but as a real force that could one day unshackle him.

Reflection also forced him to confront the bitterness that festered beneath his skin—the anger aimed at those who had betrayed or abandoned him, the rage against the society that seemed indifferent to his pain, the resentment toward himself for choices that led him here. In the cold solitude of his cell, he recognized that these emotions could either corrode him or fuel transformation. The process was excruciating—acknowledging pain without allowing it to harden into hatred, letting go of blame without erasing the lessons learned. Slowly, he reached moments of fragile grace, the rare inner peace that came when he allowed himself to embrace vulnerability rather than fend it off like a weapon. Through this, he understood that healing was not linear but turbulent, that it required confronting the very darkness he most feared. To reclaim his life meant standing face-to-face with his demons—not to conquer with fury, but to invite understanding and compassion.

The stark routine of each day gradually became a rhythm breaking the chaos, offering strange measures of stability. Each moment of stillness granted him the gift of introspection, a chance to sift through the wreckage with careful hands and begin constructing something new from the ruins. He often traced with his mind the path from boyhood innocence to hardened streets, from fleeting escapes in drugs to the harsh geometry of prison cell—each step laid bare and scrutinized not to drown in self-pity but to illuminate. The prison, a place meant to punish and contain, paradoxically became a forge where resilience was tempered. He discovered that survival was as much an internal battle as it was external and that reclaiming agency meant first reclaiming self-love amid brokenness.

He also grappled with the question of identity, the way it had splintered and fractured under pressure, how the self he once knew had been worn away by necessity and trauma. Was redemption possible for someone marked so profoundly by violence? Could he ever disentangle the tangled threads of boy and gang member, addict and soldier, lover and loner? These questions trembled on the edge of his consciousness, inviting discomfort but also possibility. He realized that to embrace the person he was becoming, he had to accept the full tapestry of his story—the light and the dark woven inseparably. This acceptance did not erase accountability but redefined it as the foundation for true transformation. His future, once a vague hope, began to take tentative form in the shadows.

There were moments when despair surged with such intensity it threatened to drown every fragile thought of hope. The weight of confinement pressed not only on his body but on the spirit, testing every ounce of determination. In those bleak instances, he returned to memories of the street—the faces of friends lost to violence or addiction, the heartbreak of family broken by absence and pain. Yet even in that darkness, a stubborn ember refused extinction. He clung to the slim promise that if he could endure these years in this cage—both physical and mental—he might someday step into a world reshaped by his own hands, free from the chains of past mistakes. This vision became a quiet beacon, a glimmer that pushed him through the endless cycles of reflection and self-reckoning.

The act of reflection proved a double-edged sword—sometimes piercing like broken glass, other times soothing like balm. Each recollection carried the risk of self-judgment, but also the potential for insight. He learned to embrace vulnerability not as weakness but as courage, to allow himself the discomfort of confronting raw emotions without flinching. In the solitude of confinement, he came to grasp that healing was less about escape and more about endurance—a painstaking apprenticeship in patience and forgiveness, grounded in the knowledge that the past could neither be changed nor erased but could be understood, owned, and transmuted into wisdom. The silence around him was no longer a prison alone but a crucible, shaping a more resilient and empathetic self, forged in the fires of isolation and reflection.

With time, he sought to connect the fragments of his fractured identity through this internal dialogue—a deep, sometimes painful conversation with himself. He asked who he was beyond the labels thrust upon him by society, beyond the roles forced by circumstance. He questioned what it meant to be worthy of love and forgiveness after a trail marked by violence and addiction. These inquiries opened the door to a tenderness long buried beneath layers of cynicism and fear. Though scars remained, they began to feel less like chains and more like testaments to survival. That recognition was perhaps the greatest achievement of all—the ability to hold his brokenness not as a source of shame but as a foundation for rebuilding.

Isolation in prison, while brutal and unforgiving, became the unnatural womb for this profound transformation. It was a place where every quiet moment forced him to confront the raw essence of his existence, where the distractions of the outside world fell away to reveal the unvarnished core of identity and humanity. The reflection he underwent was uncomfortable, relentless, but necessary. It demanded he strip away every illusion and bravado until the face that stared back from the shadows was unmistakably his own—flawed, frightened, but fiercely alive. In this confrontation, amidst the echoes of regret and the whispers of hope, he found the first fragile threads of redemption beginning to weave through the fabric of his wounded soul.

Survival Strategies

Prison was nothing like the streets, though on the surface, it seemed to hold the same unforgiving rituals and unwritten codes that determined who lived and who didn't. When the heavy metal door clanged shut behind me, sealing me off from the world I thought I knew, the reality of confinement hit me with a weight far beyond any courtroom sentence. It wasn't just the loss of freedom that strangled the spirit; it was the sudden shift from chaotic streets to the cold, calculated choreography of survival inside concrete walls. At first, every breath in the stale, recycled air felt like a reminder of failure, a punishment made manifest. But quickly, I learned that prison wasn't merely about enduring time—it was about mastering a new kind of existence where vulnerability was a luxury no one could afford.

Navigating prison politics demanded a level of awareness and adaptability I had never quite exercised before. Back on the streets, the rules were brutal but somewhat flexible, shaped by reputation, muscle, and alliances. In prison, those frameworks became rigid and almost tribal. There were factions and loyalties that ran deeper than blood, binding people in survivalist networks forged by necessity and fear. Walking the halls, I sensed the undercurrents of power and rivalry vibrating through every conversation, every sideways glance. There was a silent language spoken through posture and timing—who you looked at, how you swallowed words, the quick tightening of fists. To the untrained eye, the prison yard might seem monotonous, but beneath the surface lay a battleground just as fierce and unpredictable as the open streets at night. And I was just another newcomer, a kid who'd stumbled too hard into a world where missteps spelled disaster.

The first rule became simple yet brutal: do not show weakness. The moment you let your guard down was the moment someone else stepped in to take something more than just your dignity. I learned to straighten my shoulders and meet eyes without flinching, to carry myself as though I belonged even when every instinct screamed that I did not. Conversations had to be measured, laughter controlled, and questions carefully posed to avoid triggering suspicion. Trust was a currency traded sparingly, and betrayal was almost a rite of passage. I found myself retreating into a quiet vigilance, observing others with a blend of fear and curiosity, trying to read their intent beneath the tough exterior. My past choices played like a relentless soundtrack in my mind—how had I gotten here? What dumb decision after another had led to this cage? Each thought brought with it a pang of regret, but also an eerie clarity. Prison stripped away all illusions, forcing me to face myself stark naked, beyond the bravado and survival tactics.

Yet, amid the harshness, I discovered a strange rhythm to the days. The routines were unyielding: the wake-up bells, the clamorous calls to meals, the long stretches of boredom punctuated by moments of tension. These rhythms became both a curse and a lifeline. In the moments when the fortress of my thoughts threatened to crumble under isolation, I clung to memories—the flicker of a smile exchanged once, the feeling of sun on my skin before dawn, the promise I had made to myself that this wouldn't be the end of my story. The solitude forced me to wrestle with demons I'd long ignored: anger, fear, shame. Sometimes, the silence was so loud, so suffocating, that I felt like it would crush me. But other times, it granted me a brutal kind of clarity. I began to understand that survival was as much an internal battle as it was external. Keeping my sanity intact meant finding small corners of peace in a world designed to break you down.

In the yard, where the air tasted of sweat and desperation, I watched alliances form and dissolve like shifting sand. Some men were soldiers in their own right—imposing figures who commanded respect through sheer presence. Others wielded power with words and wounds inflicted in whispered deals behind locked doors. I quickly grasped that aligning with the wrong group could cost more than just social standing; it could invite violence that spiraled beyond immediate reprisal. Yet, remaining isolated was a precarious choice, too. Loners were targets—seen as weak, easy marks for those hungry to assert dominance. So, I learned to be cautious but engaging, striking balances between visibility and invisibility. Small acts of kindness—offering a bit of contraband or sharing food—could buy temporary shields. Observing, learning, and adapting became my daily grind.

The prison staff added another layer to the complex ecosystem. Some guards were indifferent, going through the motions; others exuded latent hostility that bled into every interaction. At times, I felt less like a person and more like a specimen under constant surveillance, dissected through hollow stares and curt commands. The arbitrary exercise of power by those meant to enforce rules underscored the daily precariousness of my existence. There were moments when frustration boiled into rage, but experience mandated restraint. Explosions of anger only tightened the invisible chains, adding more days to sentences, or landing men in solitary confinement—an abyss where sanity could unravel in a matter of hours. Instead, I focused on cultivating patience, a weapon stronger than fists in this arena of endless control.

Reflecting on my past choices became both a punishment and a source of strength. I replayed scenes from the street—the rapid escalation of drug use, the fleeting camaraderie among gang members, the moments when I turned away from family and potential paths that might have led elsewhere. Each recollection was a jagged shard in my memory, cutting deep but also illuminating how I had been caught in cycles of scarcity and survival. A part of me wanted to scream at the younger version of myself, to shout warnings and plead for change. But prison taught me that rage without purpose only consumed energy better spent on forward motion. Acceptance, hard as it was, formed the foundation for rebuilding. I promised myself I would use this time to learn, to understand the forces that had shaped me, and to prepare for what would come beyond these walls.

Books became one of my few sanctuaries. The prison library was meager, shelves holding more dust than volumes, but within those pages, I found windows out of confinement, glimpses of freedom beyond flesh and stone. Some nights, illuminated by the dim glow of a distant light bulb, I devoured stories of survival and hope, of men and women who had clawed their way back from desolation. The words seeped into my consciousness, knitting themselves into a fragile web of purpose. Therapy was a luxury few inmates accessed, but I sought out counseling groups when I could. Sitting in circles with men who carried their own scars created moments of connection that softened the edges of isolation. Vulnerability, once unthinkable, grew tentatively, stitching together the torn fabric of my identity.

Still, the pull of the streets lingered like a shadow. News from the outside was scant—snatches of gossip filtered through whispered conversations or coded notes slipped between hands. Some friends had fallen further into addiction; others had been lost to violence. My absence was a wound on my community, and I was a ghost slipping through its veins. The contrast between the brutal safety inside the walls and the lethal chaos outside became starkly clear. Prison was a crucible, a place where the hardest lessons were burned into the soul. Learning how to live without the adrenaline of the streets, without the immediate threat, but also without the false security of gang ties, forced me to invent new survival strategies. It was about endurance, yes, but also about transformation—revising the parts of myself forged in fire into something capable of healing.

I became adept at reading not only people but time itself—the ebbs and flows of moods, the cycles of threat and truce, the subtle shifts in power that could erupt into violence without warning. Survival was a dance of vigilance and restraint. I learned when to speak and when silence was safer, how to move through crowded corridors as if I were part of the background noise, how to mask fear with humor or an indifferent stare. The exhaustion of constant alertness was a wage paid to keep a fragile equilibrium intact. There were nights when sleep was elusive, haunted by nightmares reliving the streets, flashbacks of shots fired, screams, and the choking grip of addiction. In those moments, I clutched at memories of better days, held onto slivers of hope with trembling hands.

Social hierarchies were merciless, but I also discovered unexpected pockets of solidarity. Some men, despite their rough exteriors, showed kindness in small gestures—a shared cigarette, a reassuring nod, an understanding glance. These acts were the lifelines that prevented me from being swallowed entirely by despair. Religion, for some, offered a refuge, and I found myself drawn to the quiet corners where prayers mingled with whispered confessions. It wasn't so much about doctrine as about a shared humanity, a recognition that we were all searching for salvation in our own fractured ways.

Maintaining sanity became a matter of rituals and mental fortresses. I adopted practices to keep my mind sharp and my spirit intact: writing in tattered notebooks, reciting poetry under my breath, creating maps and plans for life beyond these gates. The act of creation, no matter how small, affirmed my existence beyond mere survival. I fought against the creeping numbness, against the temptation to surrender to apathy or rage. Sometimes, the weight of confinement pressed so hard I felt like I might shatter, but then I would remind myself that each day survived was a victory, a building block for the person I still hoped to become.

In this crucible, I was forced to confront who I had been and who I wanted to be. The survival strategies I adopted were more than just physical—though self-defense and caution were paramount—they were deeply psychological. I had to dismantle the identities formed by addiction and violence, strip away the defenses that protected wounds but also trapped me in cycles of dysfunction. This process was neither linear nor easy; it was a simmering, grinding effort marked by setbacks and revelations. Each small victory—the choice to walk away from conflict, the decision to listen rather than lash out—became a stepping stone toward a life reclaimed.

Inside the walls, the passage of time was warped. Days melded into weeks, weeks into months, yet moments of clarity punctured the monotony, offering glimpses of a horizon beyond incarceration. I clung to these like beacons, reminders that the harsh lessons of prison were not endpoints but parts of a longer, ongoing journey. The survival strategies molded by necessity transformed, over time, into tools for healing and growth. I began to envision a future where my past did not define me, where I could step out of the shadows cast by addiction, violence, and loss.

Prison life was brutal, unforgiving, and often dehumanizing, yet it also became a crucible for transformation. The physical confines mirrored the emotional and psychological battles raging inside me. Navigating the politics of the yard, maintaining vigilance against threats, managing the undercurrents of fear and anger—all demanded a resilience deeper than muscle or willpower. Survival was a complex dance of adaptation, connection, and self-preservation, but ultimately, it was about reclaiming a shattered self. Through every narrow escape and quiet moment of reflection, I stitched together the fragmented pieces of my story, determined that when the gates finally opened, I would step out not as a prisoner of my past but as a man still capable of redemption and hope.

Glimmers of Hope

The cold clang of the steel door echoed through the narrow hallways, a relentless reminder that my world was now confined within these walls. Prison life was a relentless conductor of time, pacing each day with the solemn rhythm of routine, despair, and the occasional flicker of something fragile, something I began to call hope. At first, every sunrise announced nothing but another day shackled by iron bars and questions—questions about choices made, roads taken, and the blurry edges of a past filled with reckless decisions and broken promises. The crack epidemic, the gangs, the violence that once defined my survival — all seemed like distant, harsh shadows now, yet the weight of those chapters pressed heavily against my chest, suffocating in their presence.

I found myself endlessly replaying the moments that led me here, trying to dissect where the fracture lines in my life had appeared. Was it the streets that consumed me? Or was it me who let those streets win? The faces of people I had hurt, or who had turned their backs on me, haunted my thoughts. The relentless noise inside the prison wasn't just the clamor of footsteps and shouted commands—it was the silence of the mind wrestling with itself. Each day behind bars stripped more layers of denial, exposing raw truths that were impossible to escape. To the outside world, maybe I was just another statistic, but inside, I was an unfinished story scribbled in the margins, desperate to be rewritten.

The challenge of confinement wasn't only the physical absence of freedom; it was the mental imprisonment that stripped away identity piece by piece. My body moved within a restricted space, but my thoughts were both a sanctuary and a battlefield. At times, the weight of hopelessness threatened to pull me into a bottomless pit. But in those fractured moments, between despair and the grinding monotony, small flickers of something new began to ignite. They were not blinding lights but gentle glimmers—whispers of possibilities beyond these walls. I started planning, quietly at first, imagining a future that I had never dared to dream of when my world was trapped in chaos and addiction.

Cellmates came and went, their stories etched into the peeling paint and graffiti on the cell walls. Some were broken, bitter reflections of the street life that had swallowed us all. Others, surprisingly, carried within them a spark that refused to be extinguished. I began to listen to their stories with fresh ears—not just the tales of crime and regret but also those of survival, resilience, and change. Slowly, I realized that redemption wasn't a myth told by the distant world outside prison but a tangible, if difficult, possibility for those willing to confront their demons head-on.

In the quiet hours, I penned down my hopes and fears. Paper became my confidant—where I scribbled plans for life outside: attending therapy, finding steady work, reconnecting with family, maybe even helping those caught in the same cycle that nearly destroyed me. The word "recovery" no longer belonged solely to the slogans hung on the common room walls; it began to take shape in my own story. Addiction, which had once held me captive tighter than any cell door ever could, suddenly seemed like a battle I could fight and win. The truth was painful—there would be no fast-track to a new life, no miraculous salvation—but for the first time, I believed in the possibility of progress, however jagged the road might be.

Reflecting on my past, I confronted the sheer magnitude of the destruction wrought not only on me but on those around me. The wishes left unspoken to my family, the friendships severed by mistrust and fear, the promises broken under the haze of intoxication. Those memories fueled a growing fire within me—a fierce desire to make amends, to not let the scars define the rest of my existence. It wasn't an easy epiphany. Shame and guilt often stormed in to sabotage my fragile dreams, whispering that I was undeserving of a second chance. But I gradually learned that hope, like strength, was both a choice and a discipline—something earned by waking up each day ready to fight for something better.

The structure imposed by prison life paradoxically became an unlikely aid in this transformation. The rigid schedules of meals, counts, and assigned duties grounded the chaos that had once ruled my days. Instead of surrendering to the numbness, I began to seek out moments of learning—books lent by other inmates, educational programs, and eventually, the mentor figure of a seasoned volunteer who saw beyond my hardened exterior. He introduced me to the idea that identity could be reshaped, that the addictions and violence I embodied were chapters but not the whole narrative. His encouragement was small but persistent, a lifeline that tethered me to the slow, painful process of healing.

Yet, even amid turning points, prison remained a place where fears festered. The shadows of violence lurked everywhere, in the sideways glances, the whispered threats, the constant vigilance that never truly vanished. Every day required navigating a delicate balancing act between maintaining dignity and survival, between embracing change and protecting the fragile seeds of hope from being trampled. There were nights marked by loneliness so intense it was almost physical, when the silence screamed louder than any prison guard. It was in those moments that the past came flooding back, dragging me into memories that I wished I could erase. But I resisted. I fought to hold onto the promise that redemption was a journey, not a destination.

Planning for the future was both exhilarating and terrifying. I dreamed of walking free, inhaling the air without the stale bitterness of confinement pressing in on my lungs. I pictured seeking out old family members whose faces had faded in the distance, hoping to rebuild bridges burned long ago. I imagined stepping into the light of a job that gave me purpose rather than survival. But beneath those hopes lay the sobering reality that the outside world had shifted while I had been locked away—technology moved forward, faces changed, society marched on without me. Reentry was a challenge I could not ignore; I mentally prepared myself to grapple with rejection, stigma, and the chaos of reclaiming a life that had unraveled.

There was a growing awareness that hope alone was not enough—action was necessary. With every passing day, I set small goals: attending group therapy sessions offered by the prison, reading personal development books, and participating in educational classes. Each effort was a brick laid in the foundation of a new self, a tangible way to resist falling back into the destructive cycles that had defined my youth. Even the simple acts of self-discipline—waking early, completing assignments, engaging with counselors—felt like victories of immense significance. These victories fed into the bigger dream of freedom, fueling an inner revolution far from the eyes of those who only saw me as a number behind bars.

The glimmers of hope that grew inside those walls were not just about personal salvation; they began to stir a deeper understanding of responsibility. I thought about the other young men still trapped in cycles similar to mine—the ones lost to addiction, violence, and despair. My story, I realized, was no longer mine alone. It carried a broader weight, a call to action that extended beyond my own survival. What could I do once I was released? How could I transform the pain I carried into something meaningful? These questions became the heartbeat of my emerging mission, a force that steadied me even when everything else wavered.

Love and connection, so long absent from my life, also found tentative roots in this time of reflection. Memories of the Canadian escort—our fragile, complicated bond—surfaced with surprising clarity. Though marred by instability and pain, that relationship had exposed a part of me willing to be vulnerable, to reach beyond walls, both physical and emotional. It was a painful reminder that redemption was not a solitary journey but one intertwined with others. It taught me that to rebuild my life, I needed to open myself to trust and healing, despite the risks.

As the days stretched into months, and months into the prospect of eventual release, I felt an emerging clarity about the kind of man I wanted to be. Not just a survivor of the crack epidemic and gang violence, or a statistic buried in a criminal record. But someone who could stand tall in his own redemption, someone whose scars testified not only to pain but to resilience and transformation. Prison had tried to break me, but in its harsh confines, I discovered the seeds of strength and a vision for a future that once seemed impossible.

The journey ahead was daunting—holding onto hope in the face of uncertainty, rebuilding trust where it had been shattered, and facing a world that might reject me wholesale. But I was learning that redemption was like a rock star's encore—not easily won, requiring relentless persistence and heart. Each day behind bars was a rehearsal, a step toward that moment under the spotlight of freedom and the life I still dreamed of living. And as the sun rose each morning beyond the concrete walls, glimmers of hope shone brighter, lighting the path toward the light I was determined to find.

Chapter 5
Seeking Structure: Military Service

A New Beginning

The day I made the decision to enlist in the military marked, in many ways, a pivotal turning point stitched into the fabric of my restless life. It was not a moment fueled by patriotic fervor or some grand vision of honor and glory; rather, it was born out of a desperate need to escape. Not just the streets that had swallowed so much of my youth, but the chaos that roiled inside me—a turbulence wrought by scars both visible and unseen. I stood at the precipice, desperate for a lifeline, a chance to rewrite the script that had played out so cruelly up to that point. The choice was mine to grasp, raw and trembling, and with it came the fragile hope that I could find structure, purpose, and perhaps some semblance of peace.

From the moment I stepped onto the military base for training, the air shifted in a way that was both alien and intoxicating. The rigid choreography of drills and the ironclad routine weren't just exercises in discipline; they were demands for conformity, a crucible forging men into something cohesive, something controlled. I learned quickly that the military did not tolerate the pasts we all carried like invisible backpacks filled with hurt and regret. Here, every second was parceled out, every emotion screened and approved or discarded, and the expectation was that you shed the skin of your old self and emerge a soldier ready to obey without question. The first days were grueling— a relentless parade of early mornings and brutal workouts that left my muscles searing and my mind teetering on exhaustion. But beneath the fatigue was a strange solace, a promise that if I could keep pace, if I could follow orders without flinching, I might finally seize something steady to lean on.

Within the confines of the barracks and beneath the loud cadence of the drill sergeants' barked commands, I felt small and exposed, but also strangely safe. The chaos of my past—the gunfire echoing in alleys, the slang and threats of gang life, the numbing burn of addiction that had tightened its grip until I nearly lost all sense of self—felt distant, like shadows flickering beyond a glass wall I could no longer reach. The military's structure carved out a space where survival wasn't about standing your ground on broken streets but about obeying hierarchies and protocols designed to impose order on disorder. Yet, even in this new world where regiment ruled, the ink of my past refused to fade quietly. Memories creeped in, uninvited and sharp—a flash of a night sky lit by gunshots, the cold sting of betrayal, the weight of loss hanging heavy on my chest. The discipline I craved battled fiercely with the jagged edges of trauma that lurked beneath the surface, reminding me that transformation wasn't a given but a titanic struggle.

Coping with these ghosts while adapting to military life tested every fiber of my resolve. PTSD wasn't a convenient label I could scribble on a diagnosis slip; it was an unrelenting storm inside me that no drill could silence. Nights were the hardest. Beneath the blackout curtains of a barracks room shared with strangers, the screams on the street outside were replaced by screams in my head—reliving moments when friends fell, when bullets pierced flesh, when trust shattered like glass. The military's stoic ethos preached strength in silence, and I learned to bury the panic, the rage, and the despair beneath a mask of calm professionalism. Yet the repression only sharpened the ache, and there were days when the tight coil inside me threatened to unravel.

Seeking help was complicated; the culture I entered valued toughness, and vulnerability was often mistaken for weakness. Still, tucked away in a small therapy room far from the clang of drills, I glimpsed the first real attempts at healing—conversations that sifted through rubble and pain, trying to build foundations of resilience.

Amid these battles, the camaraderie of newfound comrades offered a complicated balm to my wavering spirit. These men and women, many carrying burdens unseen, became both my support and my mirror. In early morning runs or late-night conversations muted by exhaustion, we shared fragments of our lives: scraps of family histories, bits of dreams deferred by hardship, moments of fear about what lay ahead. We forged bonds born of mutual survival, not unlike the alliances I had once formed on the streets, but forged now in a crucible of discipline and shared purpose instead of desperation. The paradox was not lost on me—here I was, a former gang member, learning to fight not for territory or power but for order and something greater than myself. Yet the shadows of my past followed, and occasionally the veneer of discipline cracked, revealing anger, mistrust, or the ache of abandonment.

Early in training, there were moments when the weight of my personal struggles collided violently with the strict expectations of military life. When the sergeant's voice thundered, demanding perfection, my body sometimes froze, heart racing as if I were back on a street corner facing a threat. The unyielding cadence of commands felt suffocating at times, and the impulse to rebel—so ingrained from years of living a life without boundaries—fought with the desire to conform and find peace. It was a constant tightrope walk: how to integrate the discipline drilled into me without losing the essence of who I was beneath the street-hardened skin. This conflict fueled many sleepless nights and moments of self-doubt, but it also sparked a determination to reconcile those disparate parts—to become whole not by erasing my past, but by harnessing it as a source of strength.

As weeks turned into months, the initial hope that led me to join the military transformed and complicated. The role of soldier demanded sacrifices beyond physical endurance—it required surrendering control to a system that brooked no exceptions, a surrender that was both terrifying and strangely liberating. The uniform provided a shield, a mask, a reason to hold my head high when once I had been invisible or feared. It was a declaration that I was worthy of respect, that my life held value beyond the mistakes I had made. Yet the cost was steep; the emotional cleanness I sought remained elusive, tangled in the knots of addiction's lingering grip and unresolved trauma. The military became a battlefield not just for country but for my soul's survival, laying bare the depths of my vulnerability and the height of my resilience.

The structure and predictability initially felt like salvation, but as I grew more accustomed to the rigid schedule, I noticed the flickers of restlessness ignite within me once more. The echoes of my past did not disappear; they evolved. The drill sergeant's whistle could not drown out the whispers of old cravings or the sharp sting of uncertainty about the future. Therapy sessions offered glimpses of clarity but also demanded courage I sometimes doubted I possessed. Moments of breakthrough were shadowed by relapses into old thought patterns— waking to the ache of loneliness, the fear of being misunderstood, the temptation to run from the pain instead of facing it. The duality was cruel yet real: the military promised a passage to renewal, yet also forced me to confront how far I still had to go.

Love and connection, fleeting though they were in the early days of service, began to take root as intricate threads weaving through the fabric of my new life. The fragility of opening my heart amidst such strictness threatened the walls of discipline, yet it also reminded me that vulnerability was not defeat. It was in stolen moments—late conversations under a dim lamp, the tentative exchanges that revealed pain and hope—that I learned the power of trust. I realized that healing was not a solitary journey but one that required others to walk alongside you, bearing witness to both your collapse and your triumph. This recognition did not make the path smoother, but it imbued it with meaning beyond mere survival.

Throughout the grueling days and restless nights, I clung to the possibility that somewhere beyond the drills and weariness lay the man I was meant to become. The military was more than a refuge or escape; it was a crucible where both my worst and best selves were tested, revealed, and reshaped. It demanded that I wrestle with the fragments of a broken past while charting a course toward a future touched by hope and redemption. The uniform I wore became a symbol not just of service, but of struggle—the hard-won recognition that personal transformation is neither quick nor painless, but worth every scar.

In the end, the decision to join the military was more than a shift in occupation or lifestyle; it was an act of defiance against the darkness that had threatened to consume me. It signified a refusal to be defined solely by addiction, violence, or despair. The military offered a structured lens through which to examine my own fractured identity and the chance to rebuild from ruins. Though the battle to reconcile discipline with personal demons was ongoing, and the terrain was treacherous, each day in uniform was a small victory—a testament to the endurance of the human spirit. The hope I carried at the beginning of this journey was real, fragile, and illuminated by the possibility that even from the deepest shadows, a new beginning could emerge, bright and unyielding.

Boot Camp Trials

The days at boot camp began before the sun could even think about rising, the shrill blast of the reveille tearing through the night's remnants like a blade through silk. Each morning pulled me from dreams I barely remembered back into a world sharpened to military-grade intensity. The early cold nipped at exposed skin, and the barracks echoed with the stomping shuffle of recruits fumbling through their morning routines – the clatter of boots on concrete, the rustling of uniforms being hastily donned, the muffled grunts of men trying to suppress their grogginess. But there was no room for weakness here. Discipline whispered sternly in the air like a heavy fog: stand straight, move fast, obey without question. The moments of quiet, what little they were, felt swallowed whole by the relentless, grinding rhythm that governed everything – push-ups until arms trembled and threatened revolt, run drills that turned lungs to fire, and memorization of endless codes and protocols that hammered at the edges of sanity. It was a punishing cycle designed not just to break us down but to build us into something new... something steeled and unbreakable.

Physical exhaustion became a language I spoke fluently; muscles that once remembered only the slow, lethargic motions of street survival now screamed with every movement. The drill sergeant's voice boomed like thunder, fracturing any semblance of normalcy or mercy. Somewhere in the cacophony, I found myself wrestling not only with the physical demands but also with an unseen opponent: the ghosts of my past. Each blow to my body seemed to awaken flashes of memory—corner deals, gunshots ringing into the night, the sharp sting of betrayal, the haze of crack smoke filling lungs desperate for air. There was an ironic warfare within that clashed with the discipline imposed from without. The external commands were absolute, martial in their precision. "Drop and give me twenty!" Yet the internal turmoil festered, a silent wound that no amount of push-ups could heal. I was supposed to find order here, discipline to tether me away from chaos—but inside, the battle raged on, invisible to any uniformed eye.

The mental hurdles proved as daunting as the physical. Every day demanded submersion into an environment that brooked no weakness or fatigue of the mind. Drill instructors, with their sharp eyes and sharper tongues, tore apart any display of softness like jackals stripping flesh from bone. The screaming and shouting, designed to dwarf our individuality and forge a collective identity, sometimes pushed me to the edge. I could feel the walls closing in during inspection lines, the unforgiving gaze of men who seemed equipped not only with weapons but with the ability to strip away confidence bit by bit. Some moments, standing in formation felt like standing under a microscope, every flaw illuminated, every misstep punished. It was a crucible that forced utter vulnerability—exposing insecurities, shaking the foundations built in the turmoil of my youth. I struggled to remember how to respond, how to be anything other than what they demanded. Amid the barking commands and blistering pace of training, I confronted a bitter irony: to gain control, I had to surrender control, relinquishing the survival instincts that had kept me alive on the streets, and trust this rigid structure to reshape me.

Sleep was a luxury granted in snatches, each night punctuated by sudden awakenings, heart pounding, breath shallow—not from the dream itself, but from the residue of past trauma clinging with relentless claws. Memories of gunshots and sirens would intertwine with the shrill echoes of the reveille, fusing into a confusing morass between nightmare and reality. It was during these stolen moments of restlessness that I realized that boot camp was not just about building strength. It was about the endurance of the self—the mind and spirit caught in a vise between discipline and the scars it could not erase. I felt at times as if I were splitting into two people: one a potential hewn from brute force and order, the other a fractured soul still haunted by shadows. This internal dialectic wove itself through every shouted order, every timed task, every savage drill that commanded my body while my mind drifted to darker places. There were days when it seemed the only way to survive was to block everything out, to harden into stone, numb to sensation and emotion alike—but doing so carried the risk of losing all that made me human.

The rigorous schedule forced moments of brutal self-reflection. The weight of the pack slung heavily against my back became heavier when accompanied by the burden of memories I could not shed. Each grueling march, each relentless obstacle course, was a metaphor for the long road left to walk beyond military training — a trek through pain and self-doubt toward some elusive redemption. Some recruits thrived visibly amidst the storm, their eyes bright with defiant focus, while others, like me, sometimes faltered in quiet desperation. Yet, amidst the rigidity, small glimpses of solidarity emerged from shared struggle. Bonds forged in sweat and grit revealed themselves as lifelines tethering us to humanity rather than machines. During whispered exchanges in the bathroom mirror, or short breaths shared between grueling events, camaraderie wove through the steel and grit, reminding me that these trials, though isolating in their intensity, were not endured alone.

Perhaps the most confounding aspect of boot camp was the requirement to master not only the body but also the emotions— learning to suppress fear, doubt, and anger beneath layers of discipline. Over time, a strange alchemy occurred; fear gave way to a simmering resilience born not of choice but necessity, and anger transformed into a focused edge that fueled endurance rather than destruction. I learned how to channel frustrations into movement, pain into perseverance. But this process wasn't linear. There were nights consumed by memories more vivid and raw than any physical wound—faces of those lost, moments of betrayal, the corrosive solitude of addiction's grip bleeding through the fog of war preparation. Discipline felt like both a weapon and a cage, wielding control while threatening to suffocate the parts of me that still sought understanding and compassion within myself. There was a constant battle to avoid becoming a hollowed-out shell, to cling to the fragile hope that the man I was becoming had space for both steel and soul.

The sleep deprivation, physical pain, and mental strain stacked relentlessly until moments of clarity became rare jewels in the grind. Yet, these rare moments were critical—they were the spaces where I glimpsed the possibility of transformation. A quiet evening in the barracks after lights out, when the world slowed, and the roar of orders melted into distant echoes, gave room for introspection. I thought about the boy I used to be, caught in the violent undertow of the crack streets, and the man I was fighting to become. The dichotomy between these selves felt immense. Boot camp was a cleansing fire burning away layers of past self-destruction, yet it left open wounds aching beneath the surface—a reminder that the path to redemption could not be paved with discipline alone. It demanded a confrontation with pain, shame, and vulnerability—emotions many of us were trained to bury deep beneath hardened masks.

The physical challenges I faced weren't limited to standardized drills or timed runs. There were moments when the body betrayed me—knees buckling under weight, lungs heaving in rasps that felt like they might cease, blistered hands clutching ropes that bit through calluses, and muscles trembling under the strain of carrying more than mere gear. Pushing through these limits was less about brute strength and more about mental stamina. Each obstacle wasn't just a barrier of wood or earth but a symbol of overcoming the inertia of past failures, addiction-induced apathy, and a fractured self-image. These walls we crawled up and over mirrored the invisible walls I had to scale inside myself every day. The grit demanded was not just physical but psychological—a relentless test of will against despair.

Some days the training offered brief moments of clarity and even camaraderie when shared hardships forged brotherhood. At night, lying on the hard cot beneath the fluorescent glare of the barracks' ceiling, voices softened by exhaustion traded stories of hometowns, lost loves, criminal pasts, and uncertain futures. For a fleeting second, all the shields came down, revealing broken men searching for a way out of their histories. These moments reaffirmed that beyond the rigors of military life, there was a collective yearning for redemption and belonging, an unspoken promise that discipline might be the first step toward healing the fractures addiction and violence had wrought.

Yet discipline was an unforgiving master. Mistakes were punished with brutal efficiency—extra drills, scratch lines, verbal lacerations that rippled with shame and humiliation. It was a world where errors could not be concealed or excused; accountability was immediate and absolute. I learned quickly that my past had no bearing here except as a burden I carried unnoticed, the slate wiped clean only in theory, while my body and mind bore scars that no uniform could erase. The odds of transformation felt steep. But with every bruise, every shouted reprimand, I carved out a small space where resilience gathered, where the tentative idea of a future untethered from the streets began to take shape. In this arena of pain and order, I faced the most brutal truth: there was no magic cure to erase years of brokenness. Just endless, exhausting work to become someone new, while never forgetting who I was—or what I had survived.

The weight of the training wasn't simply the burden of physical strain. It was the crushing realization that to survive this environment, I had to surrender not only my body but my very ego. Every shred of autonomy I clung to was chipped away until obedience was conditioned like a reflex, so deeply ingrained it became indistinguishable from thought. This process was both terrifying and freeing—terrifying because it meant submitting to forces far greater than myself, yet freeing because it offered the promise of structure where before had been chaos. But freedom came at a cost: internal conflict brewed fiercely as the boy who once survived by flexing defiance wrestled against the soldier who learned to obey. Sometimes, in the quiet moments after exhausting drills, I caught glimpses of the kid I left behind, the kid who still screamed for attention beneath layers of military protocol.

The trials of boot camp tested my limits in ways I had never imagined. Every hour was drenched with sweat and punctuated by moments that felt like battlefields of the psyche. I confronted the rawness of trauma, the ache of longing for a life lost to addiction, and the challenge of forging identity in the space between discipline and despair. The military demanded total submission to a new order, but the remnants of my fractured past whispered warnings and doubts that haunted my every step. Through weeks of pain, exhaustion, and mental agony, something began to shift within me—not instant redemption, but the slow forging of a resilience that felt both foreign and fiercely necessary. Boot camp was not a cure-all; it was an unyielding mirror held up to every weakness and every flicker of strength. It was hell, salvation, and rebirth all wrapped into one relentless, grueling test.

In the brutal discipline, I found moments of grace: the rising sun after a frozen morning run, the tired smiles exchanged among exhausted comrades, the rare instant when pain dipped just enough to let hope slip through. The process stripped me raw, exposed all I feared, and yet beneath the rubble, planted seeds of something new. Boot camp was more than a physical trial; it was the battlefield of self-mastery, where the hardest victory was not over enemies in the field, but over the darker parts of myself. Here, amid shouted orders and aching limbs, the journey toward healing began—fragile, incomplete, but real. Though my past was a shadow that stretched long and dark, the sweat and blood poured into each drill ached with the promise that maybe, just maybe, the boy who was lost in the crack epidemic's shadows could become the man who stood strong in the dawn light of redemption.

Battle Within

The stifling heat of the barracks was suffocating, but nothing compared to the storm that raged within me. The standard military routine hammered at my nerves with a relentless cadence: wake-up calls barked before dawn, grueling drills that left muscles trembling with exhaustion, and endless formations where silence was enforced like a weapon. Yet beneath this rigid exterior, beneath the layers of discipline and conformity, my mind twisted and turned in chaotic unrest. The battlefield outside was brutal, no doubt—but the battle within was far more ruthless, invisible to anyone around me but crushing in its relentless grip. Every ordered step through the mud and drills felt like a mask, a desperate attempt to suppress the ghosts that clawed at my sanity. My past lagged behind me like a relentless shadow, and no amount of physical toughness or mental conditioning could shake the dark memories etched deep in the fragile corners of my mind.

Military training demanded perfection, obedience, and strength, but inside my chest thumped a fragile heart weighed down by years of pain I was never equipped to confront. The crack epidemic, the gangs, violent nights where survival was a high-stakes gamble—the memories weren't past events; they were living, breathing entities that pursued me into the cold holds of the barracks. PTSD didn't announce itself with a grand entrance. It seeded silently in those moments of quiet, when the mind should have rested but instead spiraled into flashbacks of gunfire and screams, the smell of blood on the street, the cold grip of death inches away. Discipline was supposed to be my salvation, the steel framework that would forge me into someone new, stronger— but it wasn't that simple. Each drill down the endless rows of concrete bleached the color from my face a little more, and every shouted command sank deeper into my chest like a reminder: you're still trapped, still fighting a war no one else sees.

What confused me most was the paradox of control versus chaos. Military life demanded control: precise movements, complete focus, unwavering loyalty to orders. Yet my inner world was anything but orderly. Trauma had rewired my brain, setting off alarms that made trust a deadly gamble and joy a vanishing point. I wanted to be disciplined; I craved structure—the certainty of rules and routines where nothing was left to chance. But the discipline often slid into rigidity, squeezing the breath from me, pushing me toward a breaking point where the weight of old wounds pressed so heavily I couldn't tell where the man I was ended and where the scars began. The drills and commands were meant to temper chaos into something productive, yet they didn't touch the rubble inside—the raw, jagged edges of fear, anger, and shame that had become a part of me.

There were nights when the walls of the barracks seemed to close in around me. The low hum of distant voices, the shuffle of boots on pavement—it all became a dissonant soundtrack to my insomnia. My thoughts spiraled like a cyclone; memories of narrow escapes flared up suddenly, unbidden and brutal, leaving me trembling despite the physical exhaustion. I would lie awake, heart pounding, sweat soaking my uniform despite the chill in the air. The quiet became my enemy because it let the past slip into the present, blurring the line between memory and reality. I tried to squash it down with sheer will, but the war inside me wasn't something you could out-muscle. I was supposed to be a soldier, forged in discipline and endurance, but I carried a battlefield that hadn't been cleansed, a hidden front where the enemy was not a faceless fighter but my own fractured mind.

My comrades around me seemed armored in their own ways—some hardened by their own battles, others seemingly untouched by the chaos that tore through me. I envied their ability to compartmentalize, their ability to show up day after day without the invisible chains that I dragged behind me. Talking about what I felt was a luxury I couldn't afford; weakness was not tolerated here, and even if it were, how could I put into words the desolation haunting my nights? The military pride drilled into us wasn't just about physical toughness; it was about emotional stoicism, about being unbreakable. But breaking is inevitable when the pieces are already splintered.

The lines between discipline and despair began to blur. There were moments during training when I wanted to explode, to scream out the suffocating pain buried beneath my skin. I wanted to confess the terror that gripped me whenever I heard fireworks or the distant bang of a door slamming—the echoes that tore me back to dark alleyways where survival was scrap metal sharp and unforgiving. Yet, in that place, vulnerability was a danger, an invitation to be discarded or labeled. I bottled it up, locked the turmoil behind forced discipline, but it consumed me all the same, feeding on my isolation and fear. The harder I tried to impose order on my mind, the more resistant the chaos became, slipping through the cracks with relentless persistence.

My training instructors pushed me mercilessly, demanding everything—my strength, focus, and willpower—and in return, I tried to push away the memories that burned my insides. But the body remembers what the mind tries to forget. Physical exertion sometimes turned into a futile chase after peace that always seemed a step out of reach. I would run laps until my lungs screamed, sweat pouring down like a baptism I hoped might wash away the stains of addiction and violence. Yet, when the running stopped, the past came rushing back, raw and unforgiving. Discipline sharpened my reflexes but dulled my emotional awareness; the more I mastered the exterior, the more fragile I felt inside. I was a soldier forged in conflict, but the battle I fought most fiercely was within myself—a perpetual war between the discipline I had to embody and the demons I couldn't shake.

There were moments when the weight of all I had endured collapsed on me with crushing force. One night, after a brutal day of training that left my body aching and my mind frayed, I sat on the edge of my bunk and stared at the peeling paint on the wall. The silence was suffocating, and tears that I had long denied slipped down unbidden. The vulnerability was frightening, but it was also the only honest moment I'd allowed myself in a long time. I realized that the battle within was not a sign of weakness but a cry for survival, a desperate howl from a part of me that refused to surrender entirely. That quiet, painful admission was the first crack in the armor, a fissure through which hope could begin to seep.

Seeking help was another battle—a turbulent and uncomfortable step into the unknown. The military culture didn't exactly embrace discussions about mental health, and doctors were often viewed with suspicion or as obstacles rather than allies. But the crushing burden of silence pushed me to attend therapy sessions reluctantly. Behind closed doors, I confronted the raw nerve endings of my past—the faces of friends lost, nights drenched in violence, the compulsions that once enslaved me. Therapy was not a smooth path; it was jagged and painful, like tearing off a scab to reveal bleeding beneath. But it was also the first place where I felt permitted to be anything less than a rock, where discipline bent enough to allow fragility.

In those sessions, I learned to examine the tangled roots of my trauma and addiction. I realized that the structured life I sought in the military was both a refuge and a cage—a place where the external order was a fragile façade masking an internal tempest. The dichotomy between discipline and chaos wasn't a flaw; it was a conflict inherent to surviving trauma, a duality I needed to understand rather than fight. Slowly, I began to piece together the mosaic of my struggles—a story that wasn't just about defeat and despair but about endurance and the faintest glimmer of change.

As I wrestled with these invisible wounds, the camaraderie of my fellow soldiers became both a balm and a challenge. Some shared their own shadows quietly, while others remained locked behind stoic walls. Bonds formed in adversity offered moments of respite from the isolation that trauma inflicted, brief glimpses of human connection that reminded me I was not alone in the battle within. Trust was slow to build, but each shared story chipped away at my fortress of silence. The military's harsh training prepared us for external combat, but it was these moments of vulnerability that truly tested our strength.

My nights were still fraught with unrest. Dreams bled into reality in terrifying ways—gunfire echoed in peaceful sleep, faces from my past turned into nightmares. Sometimes I would wake drenched in sweat, heart hammering, gasping for air that seemed stolen. The discipline of the day could not banish the scars these dreams revealed, nor the anxiety that clawed at the edges of my consciousness. Medication was offered, a double-edged sword that dulled pain but threatened to numb the fragile strides toward healing. I was wary of becoming dependent on anything, fearing the replacement of one chain with another.

In the quiet moments between duty and sleep, I often wrestled with my identity. Who was I beneath the layers of soldier, addict, survivor, and boy raised amid violence? The military promised transformation—a rebirth into someone stronger and unbreakable— but I wasn't sure I wanted to lose the humanity that came with scars, the messy, painful truths that gave texture to my existence. Discipline was necessary, but so was compassion for the broken parts I carried. Reconciling these opposing demands became a central struggle; self-acceptance was the elusive prize I chased amid endless drills and strict routines.

There were days when rage boiled just beneath the surface, fueled by frustration with myself for how far I had come and yet how far I still had to go. The military's rigidity felt both like a lifeline and a trap. I saw others break under the pressure—some crumbling into despair or lashing out in violence—and I feared slipping into the same abyss. The battle within was relentless, a wearying war of attrition where victories were microscopic and setbacks devastating. But surrender was not an option, even when the tides of hopelessness threatened to sweep me away.

Gradually, the discipline instilled in me began to serve a higher purpose. The structure that once felt like a cage transformed into a framework that supported fragile progress. I learned to channel the chaos into determined action, to recognize when the storm inside was rising and to reach for coping tools instead of destructive impulses. This was no quick fix—it was a journey marked by false starts and painful truths—but within the hellscape of my past and military present, a tentative peace began to emerge.

The battle within was not a solitary war fought in invisible trenches; it was a crucible that shaped my resilience, teaching me that courage was not the absence of fear but the refusal to be consumed by it. Discipline became more than just a rulebook; it was a bridge between who I had been and who I could become. My wounds did not define me—they were part of a harder-earned wisdom that would one day guide me beyond the shadows. The military training, with all its pain and rigor, cracked open a door, revealing that redemption required facing the darkest corners of oneself and choosing to keep moving forward, no matter how heavy the burden.

And so, each day in that unforgiving place became a step in the long march toward healing. Each drill was a reminder that survival demanded not just physical strength but fragile courage—the courage to confront the battle within, to wrestle with trauma rather than run, to embrace vulnerability as a path to wholeness. Though the scars ran deep and the nights were often merciless, a seed of hope took root in that disciplined storm, promising a future where the ghosts of the past could finally be laid to rest.

Comrades and Conflicts

Boot camp was nothing like the streets, but off the streets' chaos and lawlessness, it carried its own brutal intensity. The moment I stepped onto that base, I was swallowed whole by a regimented world where every breath, every step, every blink was measured and accounted for. The drill sergeants barked orders like thunderclaps reverberating in your skull, snapping you out of whatever mental haze you'd carried in from your old life. For the first time, life wasn't a hustle made up in the moment, it was a sequence of predetermined movements, a suffocating routine intended to strip away any defects, weakness, or softness. It was in this crucible that I met the brothers and sisters who would come to define this raw, paradoxical phase of my existence—my fellow soldiers. Like me, many carried ghosts. Our shared traumas and fears wove a tenuous bond beneath the uniform fabric of military discipline.

Among these comrades, there was always a tension slipping between genuine camaraderie and unspoken conflict. The barracks were both refuge and battleground, a place where laughter and rage could explode under the same roof. Some nights, after the drills and humiliations, we'd gather in small clusters, voices low, sharing pieces of ourselves we barely recognized. These moments, caught between exhaustion and vulnerability, peeled back layers of posturing to reveal the pain beneath the steel. Every one of us wrestled with fragments of our broken past—some with haunted memories of streets scarred by violence, others carrying the burdens of family fractures, lost hopes, or nameless grief. We didn't always express it openly. Showing weakness was the enemy, so many masked it with cynicism or bravado, or simply shut down. But there was a deeper understanding that connected us through that thin, beaten layer of roughness.

Conflicts among soldiers weren't just about clashing personalities or the stress of grueling training. Often, they rose from the invisible weight of trauma pressing on our chests, squeezing us so tight that cracks appeared in the armor of discipline. Some comrades carried scars that manifested as violent outbursts; others retreated into silence or dark corners of the mind, unreachable by anything but pain. I remember one soldier, Marcus, whose temper flared unpredictably, turning every slight into a storm. We understood that his rage was a shield, a jagged edge hollering against the helplessness he felt inside, but it still terrified those around him, including me. I was sometimes on the receiving end of his fury, a blunt reminder that none of us was immune to breaking down. On the other hand, there was Jackson, quiet and steady, his calm surface barely concealing the nightmares that woke him at night, sweat slicking his skin. He carried his past like a secret wound, never speaking about home except in vague, monosyllabic replies. Both Marcus and Jackson illustrated how trauma was a fissure in the soldier's collective psyche—a dangerous fissure that discipline alone couldn't seal.

Despite—or perhaps because of—this volatile emotional landscape, a certain loyalty blossomed, forged in shared hardship and the relentless demand for survival. The harsh training regime required trust in each other's abilities and instincts; calls had to be obeyed instantly, and lives depended on teamwork. You learn to read the subtle signs in a comrade's body language, the shifting eyes, the barely perceptible tension in muscles. I found myself invested in others' welfare more fiercely than I'd anticipated. There was a strange kinship in this violent world, an unspoken pact to carry one another through the worst. But balancing that tenderness with the rigid military code often created internal chasms. Discipline demanded conformity and emotional suppression, yet our personal struggles required acknowledgment and, at times, reckoning. Walking that line was like balancing on a razor's edge. I often felt torn between the soldier I was supposed to become and the fractured individual still haunted by his past and uncertain future.

Military culture prizes toughness—there's no room for weakness. But weakness isn't always visible. Sometimes it's buried deep, festering beneath the intense exterior, waiting to erupt or implode. I remember lying in the tent at night, the hum of ventilators and distant machinery surrounding me like ghosts, my thoughts spiraling into dark places. Discipline tried to silence that inner voice, to stamp out the chaos of my mind, but the trauma was stubborn. It nipped at the edges of my resolve, whispering doubts, fears, and regrets. Even surrounded by my comrades, I felt isolated within the confines of my own head. Sometimes I saw the same fissures mirrored in their eyes—the fleeting shadows of anxiety or pain that discipline masked but never fully healed. We were brothers and sisters bound by the uniform, yet divided by the invisible wounds that discipline neither acknowledged nor alleviated.

Interactions among us could shift abruptly. The rage that Marcus wielded so freely was countered by moments of unexpected gentleness—a hand on your shoulder, a shared joke, a look of understanding that needed no words. Jackson's stoicism masked an intense empathy that materialized in small acts, like always making sure his bunk mates had their gear ready or quietly listening without judgment. These flickers of humanity punctuated the routine brutality of military life and hinted at the complexity beneath the surface. We wrestled daily with an identity partly imposed by the military but also forged in the crucible of our shared trauma—a contradiction of vulnerability and strength, order and chaos.

Conflict wasn't external only; it happened within me most of all. The discipline I sought as salvation was at times a stranglehold, squeezing out the very part of me that yearned for freedom. I craved the structure the military offered, a clear path away from addiction and aimlessness, yet it forced me into a rigidity that sharpened my inner turmoil. The regimen provided clarity—it was black and white, command and compliance—but it also exposed the shadows I couldn't outrun. There were moments in formation, sweat pouring down my back under the blistering sun, when I questioned everything: who I was before, what I was becoming, and if this path would lead to redemption or destruction. The drill sergeant's shouts blurred into a drone as my mind drifted, grappling with memories from my past—the gunfire, the friends lost, the nights spent begging for relief from cravings. This mental dissonance sometimes fractured my concentration, fueling frustration and conflict.

The military forged us into a unit, a family forged through hardship, yet it was undeniable that each of us carried private battles that couldn't be fully understood by those outside our personal histories. Some nights, I would walk the perimeter, staring out at the star-strewn sky, trying to find silence in the chaos of my mind. The fresh air, the gentle rustling of leaves, the distant murmur of other soldiers stirring—it all clashed with the inner noise. I thought of my old neighborhood, the crack houses burned or boarded up, the ghosts of old friends drifting like smoke through my memory. In those moments, I felt the enormity of the distance between who I was and who I hoped to become. The camaraderie—the shared laughter, the gritted teeth in unison when the drill sergeants pushed us beyond exhaustion—was a fragile lifeline amidst the turmoil.

Sometimes friction flared among us, driven not only by the relentless physical demands but by the invisible chains of our past mistakes and pain. While the training was designed to strip us of individualism, we clung fiercely to pieces of our identity, whether through small acts of defiance, moments of genuine connection, or the unrelenting pull of ingrained habits. Arguments could spark suddenly—over a careless word, a misunderstood command, or simply fatigue. The stress peeled away layers of civility, exposing raw nerves. In these confrontations, I recognized fragments of my old survival tactics from the streets: quick temper, suspicion, the need to establish dominance or boundaries. But unlike before, here there was a system to channel that energy—a consequence, a structure, a possibility for change. Even so, healing wasn't linear. The military demanded order, but chaos lingered behind my eyelids, a storm threatening to break free.

Then there was the subtle, yet powerful, struggle of reconciling my past life with this new identity. The streets had taught me one set of survival skills, the military another. In one, violence was wild and personal; in the other, it was calculated and sanctioned. The rules were different, but the tension between savagery and discipline gnawed at me incessantly. I felt the pull of old habits—the ingrained distrust, the defensive walls, the edge in my voice—colliding with the desire to belong, to be better, to protect others rather than harm. It was like trying to fit two incompatible truths into the same skin. I wrestled with the guilt of those I left behind and the shame of who I'd been, even as I pushed my body to the brink during drills, runs, and exercises. The military felt like a double-edged sword—it offered a chance at redemption, but exposure to trauma was heightened, the rules rigid, and the emotional trenches deep.

Through this inner conflict, my comrades became the closest mirrors I had. Each carried their own burden—some grief, others rage, many a cocktail of both. There was a thin line between brotherhood and rivalry, between support and competition. We saw each other on our worst days—broken, raw, vulnerable—and sometimes on our best—focused, capable, even joyful. Our shared experiences formed an invisible thread, taut with tension but never broken. We argued, clashed, and occasionally lashed out, but beneath it all there was a primitive, fierce loyalty forged in fire. It was this paradox that kept me tethered, the knowledge that even when I felt most alone, I was part of something larger than myself. That larger something was sometimes as fragile as shards of glass, yet it held a promise—the possibility of transformation.

The military exposed me to a new form of authority, one that was unyielding yet paradoxically protective. Drill sergeants embodied a harshness that demanded respect but also forced a reckoning with my own limits. Some were cruel, some fair, but all drilled discipline into us with teeth bared. Their relentless push forged an unforgiving environment where any sign of weakness risked rejection or punishment. Under their gaze, I learned to suppress the chaos inside, but that suppression was never total. The fire beneath remained, simmering and dangerous. It was a constant struggle to obey orders without erasing my humanity. In this process, I came to understand that discipline alone wasn't enough. Healing demanded more—a willingness to confront my trauma, a capacity for vulnerability, and the courage to seek help.

But asking for help was a foreign concept in that world. We were taught to rely on ourselves, to be soldiers first, individuals second. Yet the cracks showed, especially at night when silence pressed around us like a vice. I remember one dark evening when I approached Jackson quietly under the stars. The moon was a thin sliver above us, the air chill but still. For the first time, we began to share these hidden parts of ourselves without the filter of bravado or command. It wasn't a neat conversation—it was stammered, fragmented, and raw. I spoke of addiction, of violence, of feeling like a ghost both in and out of uniform. He confessed his nightmares, his fear of losing himself in the relentless machine. In that conversation, I realized that these internal struggles were the real battles we faced—far beyond push-ups, long marches, or drills. And it was in these fragile moments of connection that healing sowed its earliest seeds.

What made the conflicts bearable, the friction survivable, was the knowledge that we were all broken in ways hard to see. Respect came not just from physical strength or discipline, but from the unspoken acknowledgment that each of us carried wounds no uniform could hide. Our collective trauma was an invisible scar that unified and complicated us simultaneously. I learned to soften toward the flaws, the anger, and the silence in my comrades, understanding they were reflections of the same wounds festering inside me. We formed a fractured circle, imperfect and vulnerable, yet holding on.

Toward the end of training, something shifted in me. The discipline that had once felt like a cage began to feel, in small doses, like a lifeline. The rigid commands transformed from chains into structure, the bedrock from which to start rebuilding. I was still flawed, still haunted, but I held onto the fragile sense of solidarity that came from fighting alongside others who knew darkness intimately. The conflicts never fully disappeared—they morphed into something less jagged, more like the ebb and flow of tides. I realized that camaraderie wasn't about perfection but about acceptance amid imperfection. We were comrades, yes, but also warriors fighting battles inside and out, striving for survival and something better.

Still, the journey was far from over. The conflicts within me, between discipline and pain, security and chaos, would resurface in sudden, unpredictable ways—every new challenge in the military, every confrontation or quiet hour, rekindled the old fires. The uniform was a mask, sometimes a shield, but it could never completely hide the turmoil beneath. The burden of reconciling my past with the demands of this new life was a weight I carried daily.

Yet, through the chaos of comrades and conflicts, I began to glimpse the shape of redemption. Not as a blinding light or simple fix, but as a slow, painful climb where every step forward was hard-won. My brothers and sisters in uniform showed me that healing wasn't solitary; it was a collective struggle as fraught as it was necessary. And within that tangled web of discipline, trauma, love, and conflict, I started to believe that maybe, just maybe, there was a way out of the darkness—and a path toward becoming more than a product of my broken past.

Chapter 6
Love and Fragility

Crossing Paths

The night had that heavy, humid weight that sticks to your skin and gnaws at your patience. The city buzzed with the restless energy of people trying to forget or escape, each carrying their own invisible scars, their own desperate stories. I was drifting through those streets like a ghost, feet numb from the cold concrete, the numbness not from the chill but from a weariness that had settled deep inside me, rooted somewhere beneath decades of chaos and bad decisions. The alley where I ended up was narrow, shadowed, a place where light barely made it, but it smelled faintly of something sweet—perfume mixed with faint hints of cigarette smoke and something electric, a tang of impending change I couldn't quite name yet. In the murky quiet, I noticed her first by the way she stood—defiant and fragile all at once, a contradiction pressed into the crease of the world. She wasn't just a silhouette in the dark; she was the kind of presence that pulled the air tighter, like the calm before a storm.

Her eyes caught mine without hesitation, sharp and searching, a quiet invitation wrapped in guarded vulnerability. She was young, but there was an old sorrow lurking beneath her skin, like the story she wasn't ready to tell me yet. Her lips quivered just slightly, enough that I saw vulnerability without weakness, strength laced with cracks. We were both perched on the edges of brokenness, and maybe that's why we gravitated toward each other without the usual jaded caution that kept people like us apart. Our first words were clipped, uncertain— her voice soft but with an edge as if she'd been forced to harden herself time and again. I can't remember what exactly sparked that awkward first conversation, maybe a shared cigarette, the broken rhythm of two lives out of sync with the world around them, but I do remember that electric tug, the feeling that meeting her could rewrite everything I thought I knew about survival, about trust, about love.

The city kept humming behind us, indifferent to our small bubble of connection. We talked in half-truths, guarded revelations punctuated by nervous laughs and occasional silences that stretched longer than comfortable. She had stories—damaged, tangled stories about being a Canadian escort, living on the fringe of acceptance, in a world that looked at her with a mixture of disdain and want. But beneath that tough exterior was a heart that seemed to tremble with the same desperation I knew all too well. Our worlds, so different yet so parallel, began unfolding slowly in the dark alleyway where we met by chance. We spoke of loneliness that felt suffocating, of dreams we had buried under layers of bad choices and survival instincts turned too sharp by pain. The barriers we put up between us were fragile, crumbling under the weight of shared understanding. We weren't just two lost souls passing in the night; something more precarious, more fragile was beginning to grow from the ashes of our battered lives.

But it wasn't all tenderness—no, the reality of who we were and what we carried clung to us like a second skin. Her profession meant a world rife with danger, unpredictability, and emotional walls that were hard to breach. I carried the ghosts of my own battles—gang affiliations, addiction, wounds that healed unevenly—telling me that love was a luxury we couldn't afford. There was fear in the way we both edged around the idea of closeness; the weight of our histories made intimacy a risky gamble, like trying to hold onto smoke before it vanished. Every tender glance was haunted by the uncertainty of what tomorrow might bring: violence, betrayal, or just the cold indifference of a world that never gave us a fair shot. Yet in this mix of yearning and fear, I found myself drawn to her more fiercely than I'd expected, wrapped in the dangerous hope that maybe, just maybe, we could carve out a moment of peace amid the chaos.

The city was a backdrop full of cracks and shadows, but our connection cast a fragile light that flickered stubbornly—sometimes steady, sometimes flickering—against the oppressive dark. We began seeing each other more, stolen moments between the rubble of our lives, a delicate dance of holding on and letting go. Each encounter was charged with a blend of raw emotion and the unspoken understanding that this relationship could either save us or unravel us completely. There were moments bathed in warmth—her head resting on my shoulder, a rare stillness in a world in constant motion—but also moments fraught with tension, as the past echoed loudly, reminding us why trust was a battlefield strewn with traps. The instability of our lives pressed in hard: nights where she vanished without warning, days when I was pulled back into old habits, the constant struggle to believe that two broken people could build something real without falling apart.

Yet, despite the chaos, that fragile connection was a thread weaving its way through the jagged edges of our existence. Sometimes I'd catch glimpses of softness in her laughter, a fleeting break from the hardened shield she wore like armor. I started to see beyond the tough exterior to the person underneath—a woman craving authenticity yet afraid of exposure, a survivor wrestling with demons that mirrored my own. Our conversations deepened, moving from surface-level exchanges to raw admissions, peeling back layers of pain and resilience. We shared dreams that shimmered briefly, ephemeral but potent— visions of a life unchained from addiction and fear, a life where healing could take root. But each hopeful thought came with an undertow of doubt, because in our worlds, hope was a fragile thing, often crushed under the weight of reality's relentless grind.

As we tangled in the complexities of our bond, the difficulties of our unstable circumstances loomed large. Money was scarce, survival instincts razor-sharp; danger was never far behind. Her world was punctuated by encounters with clients who blurred the lines between protection and predation, moments that left scars I could never fully reach, but saw reflected in the raw edges of her gaze. My past was equally unforgiving—gang conflicts simmering beneath the surface, the constant battle to stay clean, to keep a hold on some semblance of self amidst chaos. We were two wounded people in a precarious dance, trying to hold space for tenderness while the shadows of our past threatened to consume everything. Arguments flared, fueled as much by pain as by fear—fear of being abandoned, fear of being hurt again, fear that this fragile connection was nothing more than a mirage. But somehow, beneath the volatile surface, there was a thread of something profoundly real, a possibility for redemption that neither of us clearly understood, yet were desperate to grasp.

It was in those moments of conflict and reconciliation that I realized how deeply intertwined pain and hope could be. We weren't just surviving anymore; we were beginning to live in a way neither one of us had dared before. The cracks in our armor revealed vulnerability, and with it, the chance to heal. We tried to protect each other from the harshest truths, but the truth found its way in regardless, demanding honesty, demanding growth. Our nights together were bittersweet— filled with stolen kisses and whispered confessions that dared to dream of stability, even as the world outside threatened to tear everything apart. There was no neat resolution, no fairy tale endings here, just two people leaping towards each other in a storm of doubt and desire, desperate to find a foothold in shifting sands.

Looking back, that first crossing of paths between us was more than just chance—it was the beginning of a turbulent journey marked by raw confrontation and fragile tenderness. It introduced me to a kind of love that didn't erase pain but held it close, letting it breathe beside hope. Amidst the broken streets and broken lives, she became a mirror reflecting my own struggle and a beacon shining faintly in the distance. The relationship was never simple, never easy, but it was real—a living testament to the possibility of connection when everything else seemed lost to darkness. And in that, even the rough edges, the inevitable heartbreaks, and the unstable moments carried their own kind of beauty, reminding me that even from the deepest shadows, we can find sparks of light if we dare to look for them.

Unlikely Bond

The night was thick and heavy with the kind of silence that clings to the bones, the city's distant hum just a low murmur beyond the cracked windowpane. He sat alone on the thin mattress pushed against the crooked wall of the cheap apartment, the flickering orange glow from a dying streetlamp casting jittery shadows that played tricks with his tired eyes. In that fractured space—something between loneliness and despair—the unexpected arrived, soft-footed and deliberately unnoticed at first. She was a fragment of a different world, someone who didn't exactly belong in the gritty landscape that had held him captive for so long, yet she walked right into it without hesitation. Their worlds collided like conflicting thunders, and from that collision came an improbable tenderness neither could have predicted.

When they met, it was in the kind of place where everything lined up against any notion of safety—an all-night diner smeared with grease and neon, the kind of joint people frequented not for comfort but for the cheap anonymity it offered. The room smelled of burnt coffee, stale cigarettes, and the faint tang of desperation that seemed permanently embedded in the cracked vinyl booths. She sat alone, fingers tapping nervously on the table, eyes scanning beyond the walls as though trying to find a way out of a loop she had no idea how to break. There was something fragile in her posture, like a bird wounded yet stubbornly clinging to flight. When he approached, it felt less like a stranger crossing his path and more like a broken soul reaching for another to lean on.

Their first words were clipped, rough-edged and punctuated by uneasy silences, but seeped with a mutual recognition—two people broken by vastly different battles, yet bound by the silent scream of loneliness only those who have known real loss can hear. She wasn't the kind of person who walked the streets unnoticed, but there was no glamour in her presence here, just a misplaced vulnerability wrapped in carefully folded layers of guardedness. He saw past those layers, those walls she didn't even realize she was building anymore, to the woman who beneath every fragment of pain still longed for connection. In her, he recognized the same flicker of survival that had kept him clinging to life during those darkest nights. It was raw and chaotic, but somehow it was real.

Together, they found a strange rhythm in chaos, like two outcasts scribbling new lines into an already crumpled story. Their conversations unfolded slowly, unhurried, a mix of laughter that was foreign and uncomfortable, and staggering confessions that felt heavy on the tongue but lightened the heart just enough to keep going. They shared pieces of history that fit awkwardly—the broken homes, the moments when addiction and desperation had cornered them, the nights when survival felt more like surrender. It wasn't the kind of connection scripted in the romance novels or the movies, but it was genuine, birthed from the kind of pain only the shattered know how to cradle. He found himself caring for her in ways he never thought possible, not despite their broken pasts, but because of them.

Her presence was a mirror reflecting his own struggles; she was both company and confrontation—a living contradiction who challenged the blurry edges of his world. He could see it in the way her eyes danced when she caught glimpses of hope, or how they hardened in moments of despair. They didn't pretend to understand each other fully, and perhaps that was the point; the unspoken acknowledgment that sometimes love wasn't about saving, but about simply holding space. In the moments where words failed, a touch or a glance offered solace stronger than any promise. It was messy, unbalanced, and raw—neither of them able to steady the other but somehow steadying each other all the same.

Yet, beneath this fragile bloom of intimacy, a storm brewed relentlessly. Their lives, tethered by instability and haunted pasts, refused to bend quietly to the will of affection. Every step forward was shadowed by mistrust, ghosts of old wounds lurking in the corners of their interactions. She carried scars that whispered of exploitation and survival on cold nights fifty miles north on unfamiliar streets, scars that didn't fade beneath the soft gaze of someone new but instead flickered defiantly in the dark. He wrestled with the demons of addiction that pounded at the door every time they tried to build something real, the clutch of old habits threatening to unravel the threads they carefully wove together.

The city's undercurrents were never far, reminding them of the precarious tightrope they walked. Friends from his past, some still caught in the spiral of gangs and substances, hovered around edges that tasted like old mistakes. Her own world was filled with blurred lines— a profession born from necessity rather than choice, exposing her to dangers that unsettled his newfound tenderness. They lived on the precipice of chaos, caught between yearning for normalcy and the inescapable gravity that pulled them back into survival mode. There was a constant tension, an earthquake beneath the surface, that made every moment with her feel like skating on thin ice in a storm.

Still, despite the turmoil, the intimacy between them grew in ways neither foresaw. It was built on honesty, fierce and unfiltered, terrifying in its demands for vulnerability and trust. She challenged his self-imposed walls, forcing him to confront the parts of himself he'd buried deep beneath layers of pain and rage. In turn, he became a rare constant in a world that offered her little consistency—a witness to her struggles, fears, and fleeting moments of joy. They navigated the treacherous dance of love and survival—sometimes stepping too close to the fire, sometimes retreating into shadows where silence reigned. The cracks in both their defenses were widening and mingling, creating a fragile mosaic of hope that refused to shatter completely.

Her laughter, when it came, was like a jagged burst of light cutting through months of darkness. Those moments, as brief as they were, breathed life into the stale air of his apartment and into his battered soul. In her eyes, he glimpsed possibilities beyond the broken streets, beyond the violence and addiction that had defined him for so long. She believed in the potential of change, even if she doubted it could ever fully reach her, and that belief became a lifeline for him. They were not saviors to each other, far from it, but their broken pieces fit together in a way that made the jagged edges bearable.

Yet, love in their world was never a sanctuary free from the harsh realities clawing at their backs. Nights were spent awake, shadows creeping over memories too painful to face. Arguments flared, fueled by frustration and a deep-seated fear of losing what little they had. They wrestled with jealousy, insecurity, and the perpetual question of whether survival alone was enough to hold two fractured souls together without tearing each other apart. Trust was a tentative thread, frayed by past betrayals and the unforgiving weight of their circumstances. Many nights, they lay in silence, hearts pounding with the unvoiced truths that threatened to suffocate the fragile bond.

Compounding their struggles was the fact that neither was truly free. He was haunted by a past that refused to loosen its grip— incarceration's shadow, the rigid structure of the military still echoing in his mind, the lingering pull of addiction that whispered seductively in moments of weakness. His battles were internal wars waged in quiet desperation, his steps toward recovery vacillating between triumph and relapse. She fought her own relentless demons: the stigma of her choices, the danger that stalked her profession, the loneliness bleeding into every conversation. Together, they existed in a constant state of flux, clinging to the hope that love might be the key to break their cycles even when the world conspired otherwise.

Their relationship was a study in contradictions—a fierce tenderness mingled with fragility, a passionate connection clouded by insecurity, fleeting moments of peace threatened by the inevitability of chaos. But in the midst of this tumult, it was also something rare and beautiful: a testament to resilience. Two people unwilling to give up on each other, even when the odds stacked so heavily against them that giving up seemed the easier choice. The nights they spent sharing whispered dreams of a different future were acts of rebellion against the harshness outside their door, a stubborn stand against the darkness that had tried to claim them for so long.

This unlikely bond became a lifeline, fragile but real, a small island in the floods of despair. They navigated the broken landscape of their lives hand in hand, learning slowly that vulnerability was not weakness but strength, that love could be both chaos and calm. In her, he found a reflection of his own fractured humanity—a reminder that no matter how deep the wounds, the heart still held capacity to heal, to hope, and to love fiercely against all odds. And she, too, found in him a glimpse of grace, a chance to rewrite the painful narratives that had shaped her existence.

Their journey was far from perfect. There were setbacks, nights when the past threatened to swallow them whole, moments when the cracks in their armor let in more darkness than light. But through every trial, they chose to keep reaching for each other, holding onto the fragile thread that connected two battered souls amidst a world that often felt determined to tear them apart. In the fragile space between despair and hope, they discovered a shared truth: that love, in its most untamed and imperfect form, could be the spark that ignites the slow, stubborn flame of healing and redemption.

Turbulence

The night it all spiraled in a storm of chaos and fragile hope, I found myself caught on the ragged edge between desperation and something I barely dared call yearning. She appeared in my life like a sudden flash of light — a Canadian escort with eyes full of secrets and sorrow that mirrored mine, reflected back through a cracked, half-broken mirror. There was an immediate recognition, an unspoken admission of battles fought in silence, invisible wounds shaped by the world's merciless grip. Meeting her wasn't some grand cinematic moment; it was the collision of two vulnerable souls in the dim, grimy corners of a city that swallowed people whole. A fleeting touch, a stray glance, the careless brush of hands, and already, the weight of our own personal demons settled between us like thick smoke, suffocating yet oddly comforting.

From the beginning, our bond was forged less in joy and lightness than in shared turbulence. She carried her own scars, draped beneath the lacquered exterior of her sharp smiles and practiced poise. Behind the facade, there was pain carved deep — a mixture of exploitation, self-loathing, and a desperate clutching to something real amidst the transactional haze. I, too, was drowning. Addiction had wrapped its tendrils tightly around my spirit, squeezing out hope and leaving behind a grim landscape of paranoia, rage, and aching loneliness. Together, in the flickering shadows of streetlights and motel rooms, our connection was as much about survival as it was about affection, strained and brittle yet undeniably carrying the seed of something transformative.

Our conversations were jagged, raw, and often interrupted by the crashing waves of our own internal storms. I remember sitting on cracked vinyl couches, the air thick with cigarette smoke and unspoken truths, as she spoke quietly of her past — the desperation that led her into escorting, the nights when she felt more like prey than human, and the ghost of a childhood stolen by neglect. There was a fierce intensity beneath her words, a cyclical torment that echoed my own struggles with gangs and addiction, with betrayal and loss. Sometimes, our voices would rise in frustration, anger flaring from the pressure cooker of our worlds colliding. There was no room for pretense; too much was broken, too much raw to hide behind polite smiles or shallow niceties.

I was often a mess of contradictions — craving closeness even as my addiction clawed its way to the surface, threatening to undo any semblance of trust we tried to build. She would watch me wrestle with invisible chains, the shaking hands, the jittery eyes, the moments of cold clarity that turned quickly into reckless fury. I wanted to be better for her — wanted to create a space where we could just breathe without the suffocation of our pasts — but the ghost of the streets and the relentless pull of crack's cruel embrace often sabotaged those attempts. Our relationship was a battlefield strewn with moments of tenderness and equally sharp episodes of abandonment and misunderstanding.

External pressures bore down on us like a relentless tide. Friends and associates from my old life sneered at the idea of me softening, whispering that vulnerability was weakness. Her world, too, was fraught with precariousness; the transactional nature of her work kept her constantly on the edge, battling feelings of worthlessness and the ever-present danger lurking behind every closed door. We were both juggling precarious identities — me as a man trying to claw back from addiction, her as a woman negotiating survival in a line of work that consumed privacy and self-respect alike. Each meeting was a tightrope walk between hope and despair, where one wrong step could send us tumbling into the abyss.

I often found myself haunted by jealousy and suspicion, shadows cast by the violence and mistrust that had become ingrained in my DNA. There were nights when her attractiveness drew unwanted attention, and my past's toxic masculinity flared up, manifesting as possessiveness that only pushed her away. She needed space and understanding; I could only offer jittery fragility and a desire to grip too tightly, afraid to lose her in this world that seemed designed to swallow us whole. Arguments erupted over mundane things, but underneath was a fierce struggle — fear of abandonment, shame over our failings, and the daunting weight of our own inner battles. We tried to build a fragile cage of trust, but both our wings were broken, and the slightest breeze threatened to shatter it entirely.

There were moments of breathtaking beauty amidst the chaos. Fragments of tenderness where she would cradle my head, whispering soft reassurances that felt almost alien, yet desperately needed. In those times, the world outside—the grime, the danger, the addiction— seemed to recede, leaving space for something close to peace. It was a fleeting oasis, fragile and tentative, but enough to remind us both of the humanity buried beneath the scars and pain. She wasn't just a person I kept close out of necessity or habit; she was a mirror that reflected my own brokenness, and in her, I glimpsed the possibility of redemption, though it remained elusive and distant.

But the turbulent currents were relentless. The pull of drugs was a constant shadow between us, twisting even the kindest moments into tension-filled hardships. Relapses drained our resilience, each one a punch to the gut that set off chains of blame and despair. I could see the exhaustion in her eyes—the dawning realization that love alone couldn't build the bridges we needed. Sometimes, she vanished for days, retreating into the anonymity of her world to escape the suffocating closeness of my unraveling. Other times, I would find myself spiraling down alone in the dark, haunted by guilt and the crushing weight of being unable to protect either of us from the destruction we both carried within.

Our financial desperation added another layer of complexity. I was barely holding on to any semblance of steady work, and she earned what she could on the margins, but the constant uncertainty gnawed at us both. We lived hand to mouth, in cramped, rundown apartments that reversed slowly into decay just like our own lives. Bills went unpaid; meals were skipped, and simple luxuries faded into ghostly dreams. The stress wove into our interactions, turning every discussion into potential conflict, yet surrendering to the bleakness was not an option we entertained readily. We wanted to rise above; the dream of a better life was a fragile thread we clung to, even as reality bit deep into flesh and spirit.

Within the turbulence lay a paradox — a fierce yearning for stability coupled with self-sabotaging tendencies born from years of pain and neglect. Trust was a battlefield, each hesitant step forward met with doubt and the ghosts of past betrayals. She wrestled with her own demons — self-worth battered by a world that commodified her very existence, while I was shackled to the chains addiction clamped down on me so mercilessly. Our fights often stemmed from these unmet needs, a misplaced attempt to claw for control in the uncontrollable chaos. But beneath the harsh words and tears were tentative rhythms of connection, a stubborn pulse that kept the fragile flame burning.

One night stands etched themselves painfully into memory — the hollow coldness of isolation even as we lay tangled together, bodies entwined but hearts distant. We would hold each other close, seeking warmth and comfort, but both haunted by the unspoken fears that lurked beneath. I could hear the thunder of my own heartbeat in the silence, a frantic reminder of how desperate I was to escape the wreckage of my mind. She whispered promises that she sometimes believed, and I repeated them back like mantras, our words less real than the whispered hope that clung tight despite everything.

Outside forces crashed against us like a ceaseless storm. People from my past circled like vultures — old crew members demanding loyalty, threatening if lines were crossed. The streets did not forget nor forgive, especially a man trying to find a new path after so many missteps. She bore the brunt of that reality, the worries etched into her expression on my darkest days. Jealousies flared in shadows beyond my control, distrust seeded by the raw edges of where we came from. And yet through every brutal test, she stayed, sometimes hurt and angry, but present.

Our attempts at envisioning a future sometimes lit brief fires in the gloom. We dreamed aloud of leaving the city, of a small place by the water or tucked in the woods where the air was clean and the nights silent except for honest conversation. The idea of building a life together was intoxicating, but always tangled in doubts about whether we could ever outrun the legacies selling us out to ourselves. The promises we made were fragile things—better than nothing, but easily shattered by the crashing waves of addiction and fear.

Despite the turbulence, there was a strange grace in our entanglement. It was as if the brokenness we carried became a language only we could speak, giving us both permission to be imperfect, broken, and yet still worthy of love. Somehow, the chaos made the moments of clarity brighter — a smile that wasn't forced, a touch that lingered unafraid, a glance that held a thousand unspoken apologies and hopes all at once. It was messy and difficult, sure, but real in a way few things had been in either of our lives.

The nights were the hardest — long stretches where silence spoke louder than words, punctuated by fits of restless sleep and waking nightmares. Fiery arguments would give way to hollow reconciliations before the dawn, when exhaustion forced us to retreat back into the fragile bubble where pain softened just enough to rest. I learned to read the subtle cues in her presence, the slight tremble in her hands or the guarded way she spoke, and in turn, I tried to hold myself together, to be the anchor she needed even when I felt more like a storm myself. Sometimes that was enough; sometimes it wasn't.

She brought moments of vulnerability, too — confessions whispered in the dark, tears shed without shame, glimpses of a little girl lost in the person she had become. Those moments made me fiercely protective but also painfully aware of my own limitations. I wanted to be her savior, but often I was only a reflection of the chaos we both inhabited. We were mirrors cracking each other open in ways that sometimes led to healing, but more often to bleeding wounds.

Even in the depths of our personal tempests, we found little sanctuaries — evenings spent sharing music, the gentle brush of fingertips tracing the outline of dreams yet unrealized, laughter that shattered some of the heavier shadows even if only for a fleeting heartbeat. Those shards of light kept us tethered to a shared humanity, a reminder that despite everything, we were still capable of connection.

But the external pressures were unrelenting reminders of the precariousness of what we held. Threats from old enemies, the constant pull of addiction, the weight of both our pasts — all exerting their relentless gravity. Sometimes I could see the resignation in her eyes, the quiet reckoning that love alone might not be enough to keep us afloat. And yet, she stayed, just as I did, holding tight to the fragile thread of hope that maybe, just maybe, we could carve something out of the wreckage.

Our story was not a simple one of love conquering all. It was war fought on multiple fronts — between our own self-doubt and desire for redemption, between moments of connection and the aggressive forces eager to tear us apart. It was messy and raw, beautiful and brutal, filled with the kind of contradictions that only real life can spin. I learned that turbulence was not just about external conflict but a constant internal battle, a struggle to hold on to identity and hope even as the world threatened to unravel everything.

In the end, that tumultuous relationship laid bare the depths of my own brokenness and my capacity for growth. It showed me that love's power was real but fragile, that connection required more than just desire — patience, understanding, forgiveness, and the courage to face demons both old and new. She was a storm and a calm, a mirror of pain and a glimpse of hope, and though our time together was marked by struggle and strife, it was undeniably a chapter that shaped the man I struggled to become.

The turbulence of our relationship was a crucible, shaping, burning away illusions, leaving raw truth exposed but also a chance — however frail — to reach for something beyond the wreckage. Her presence was a bittersweet reminder that amid the darkest corners of my existence, light could still flicker, fragile and flickering, inviting the possibility of redemption. It was not a perfect story, but it was undeniably ours — a testament to the complicated and beautiful mess of being human, struggling to love and be loved even when the odds were impossibly stacked against us.

Chapter 7
Homeless Nights

Lost and Alone

The nights were the hardest. The city around me morphed into a cavernous beast of cold concrete and flickering streetlights, indifferent to the flicker of life scurrying beneath its vast, uncaring silhouette. I walked the cracked sidewalks, clutching whatever thin layer of clothing I had against the biting wind that seemed to pierce straight through my bones. At times, the chill wasn't just the cold. It settled deep in the pit of my stomach, a gnawing hunger that went beyond food—an ache born from isolation, from the weight of invisibility pressing on me harder than the rough tarp I had wrapped around myself in those desperate moments. Losing my home wasn't just a physical displacement; it tore through my spirit with a brutal precision that no one who hadn't lived it could readily understand. There I was, lost and alone amidst a cacophony of city noises—sirens, distant shouts, the shuffle of feet, the hum of traffic—but none of it reached beneath the surface of my solitude. The rawness of being homeless was a constant reminder of just how precarious life could become when stripped of security, love, and hope.

I remember the first night I slept on the street, the way the cold bit through every layer I had, the unforgiving texture of the sidewalk pressing against my cheek like a reminder of where I belonged now. Survival meant having to reconcile with the fact that I was utterly unprotected. My mind, once consumed with dodging gangs and trying to scrape together whatever life I could, now had to be laser-focused on finding a place that granted the barest semblance of safety.

The street was no refuge—it was a battleground where every sound could signal danger, and every shadow hid potential threats. My breath formed haphazard clouds in the freezing air, mingling with the smog and despair that clung to the city's underbelly. Night after night, I forced myself out of the fog of exhaustion and numbness, eyes darting to every corner, every alleyway. Every small victory—finding a secluded bench, a warm bus station floor, or the kindness of a stranger who handed me a cup of coffee or a granola bar—became a lifeline, but even those moments were tinged with the ache of invisibility, as if I had already been erased from society's ledger.

Loneliness was an unbearable companion. It nestled into my bones and refused to budge, gnawing at the edges of my mental fortitude. The streets had no mercy for loneliness; rather, they seemed to amplify it, turning it inward until I felt like a ghost wandering through a world that didn't notice or care. The faces that passed me by were often blurred, expressions glazed over, eyes averted. Being homeless made you less than human in the eyes of many, a judgment that weighed heavier than any cold or hunger. Yet, inside me, the memories of better days, of laughter and love, flickered like dying embers—fragile, flickering, but never fully gone. That faint light kept me from disappearing entirely, a tether to my former self, a reminder that there was something to fight for beneath the layers of grime, despair, and exhaustion.

Each day was a constant negotiation between fighting against the degradation that homelessness imposed and finding small glimmers of dignity. I scavenged through trash bins not out of shame but necessity, searching for discarded food or clothing—a practice that taught me the city's cycles of waste and want with stark clarity. I learned to recognize which shelters offered a bed for the night and which were overcrowded or unwelcoming, sometimes driven more by prejudice than policy. The streets were a brutal school, teaching lessons in humility, resilience, and the art of invisibility. I became adept at slipping through the cracks, blending into the background to avoid harassment or worse. But the isolation wasn't just physical—it seeped into my thoughts, feeding a self-doubt so potent it became almost a physical presence. How had I come to this place, where my own reflection was a stranger, shadowed and gaunt?

My resourcefulness was a double-edged sword. On one hand, it was the key to survival; on the other, it served as a painful reminder of how far I had fallen. I learned to navigate street politics, where alliances were as fleeting as the daylight and where one misstep could mean losing a safe spot or worse. I found solace in small routines—discovering open libraries during the day, sitting under their fluorescent lights to escape the street's merciless gaze, or combing through second-hand bookstores, where the silence was pure relief. Those pages became my refuge, worlds that allowed my mind to drift beyond immediate hardship, even if only briefly. Yet, the return to reality was jarring. The moment outside those walls was a plunge back into the cold, back into the roar of urban neglect.

There was a constant tension between hope and despair, a pendulum that swung violently with every passing day. Sometimes, a wave of despair crashed over me—a gnawing conviction that I had lost everything and that the streets would claim me entirely. Other times, hope flickered faintly. It arrived in the form of a sympathetic smile from a passerby, an old friend who reached out in a moment of clarity, or an overheard conversation about community programs that could offer a hand. But most often, I wrestled with the question: How could I reclaim my life when the very fabric of who I was had been so badly torn?

The nights stretched long and agonizing, yet I discovered a strange kind of solidarity among the lost. Even in the coldest, darkest hours, the street creatures—other homeless men and women, addicts, ex-cons, dreamers turned to ashes—shared unspoken bonds forged through mutual hardship. We exchanged stories in whispered tones, offering what little comfort words could provide. I learned to read faces hardened by pain but softened by fleeting compassion, to recognize the flicker of a smile despite circumstances that often left us broken. In those moments, I felt less like a nameless shadow and more human, a glimmer of connection flickering amid the darkness.

Amid this harsh landscape, I became painfully aware of the layers of trauma layering upon one another—my past, my addiction, the violence I had escaped from, and the daily beatings handed down by life on the street. It was like carrying invisible wounds that never fully healed, even as they scabbed over time. I tried to numb the pain in various ways, falling into old habits, sometimes succumbing to that fleeting, soul-crushing escape that drugs offered. But the pull of addiction was tricky; it promised relief yet dragged me deeper into the mire. Struggling to stay clean or at least minimize the damage felt like a war waging beneath my skin, with no clear victor. It's easy to romanticize survival—but the truth was raw, ugly, and often suffocating. I felt myself slipping, caught in cycles of despair punctuated by moments of clarity that were as frustrating as they were fragile.

Despite everything, I was still learning how to cope. I began to notice small acts of kindness as acts of grace. A volunteer who handed out warm meals, someone who listened without judgment, a place that welcomed me for more than just my utility. These encounters, while rare, reminded me that humanity still existed beyond the grime and desperation, that compassion could survive even in the most broken places. I started to understand that my story—though marked by failure and pain—was also one of endurance. The fact that I was still breathing, still moving forward in some small way, was itself a form of resistance.

The emotional toll was crushing. I often woke up feeling hollow, veins emptied of hope and joy, heart aching with the loneliness that no amount of alcohol—or drugs—could fill. I grappled with shame daily, a relentless voice telling me I was broken beyond repair. But with time, I found an inner voice—uncertain, hesitant—that whispered a different truth. It spoke of survival not as a shameful defeat but as an act of courage. This voice grew louder the more I leaned into therapy, the more I reached out to advocacy groups and those who had walked these paths before me. Reflecting on my experience began to give pain a shape, a purpose, something beyond pure survival. It turned into a story I could wield, a message I could send out into the world to warn, to comfort, to incite change.

My days on the streets became a crucible of self-discovery. The invisibility that once threatened to erase me forced me to confront my identity stripped bare. Who was I beneath the layers of hardship? What remnants of that boy who once dreamed of something better still lingered in the wreckage? With gritted teeth and a stubborn heart, I started piecing together a shattered sense of self. I began to realize that homelessness was not the end of my story, but a chapter—painful and dark, yes, but not final. I discovered within myself a fierce crackling spark of resilience that refused to be snuffed out.

This period of my life was defined not just by loss but by the hard-earned knowledge that even in the deepest shadows, it's possible to find a sliver of light. The nights on the cold streets taught me the meaning of endurance, of raw survival, but also drove me to seek out healing, connection, and eventually redemption. Through the desolation, I emerged with scars etched deep into my skin and soul—but those scars became the map back to life, to hope, and to the possibility of reclaiming the person I was meant to be. I was lost and alone, yes, but more importantly, I was beginning to find myself again in the fragments of that lost existence.

Street Survival

Every minute on the streets was a calculated gamble, a relentless negotiation between survival and surrender. The air itself seemed charged with a hostile energy, thick with desperation and watchful eyes, where every passerby could be a friend or a threat, and the line between the two blurred beyond recognition. Street survival—navigating this labyrinth of decay and danger—demanded more than instinct; it required a warrior's cunning, the kind you weren't born with but forged through fire. I learned early on that the streets don't care about your intentions, your dreams, or your past. They only recognize your ability to adapt, to protect what little you had, or more often, to take what was necessary without hesitation.

The day started with an unyielding chill that seeped through thin clothes, biting into skin that had long grown numb to cold and pain alike. Shelter was never guaranteed; it was a precious and fleeting commodity earned through uneasy alliances and sheer luck. Cardboard boxes, tattered blankets, and the crumbling alcoves of abandoned buildings became my fortress, but inside, safety was an illusion. Rodents skittered nearby, their tiny claws scratching against the pavement, a reminder that this patch of concrete was a contested battleground. Noise was constant—a cacophony of distant sirens, barking dogs, muffled fights, and the low hum of frantic conversations whispered among shadows. The street unfolded as a living organism, breathing threats, secrets, and the relentless pressure to stay alert. To let your guard down was to invite disaster, a lesson ingrained in me after too many betrayals in the dark.

The dangers were not always visible. Predators lurked in every corner, some obvious with the gleam of hostility in their eyes, others cloaked in familiarity, wearing the disguise of peers bound by circumstance. Trust was a currency too costly to spend frivolously. I moved through the maze with a practiced wariness, learning to read body language like an open book. The slightest twitch of a shoulder, a sudden halt in footsteps, a glance that lingered too long could signal impending violence or theft. It was a dance choreographed by necessity—when to look away, when to stare down aggression, when to fade into the background. Every interaction was a test, every pause a calculation. In this world, your reputation whispered loudly on the wind, determining whether someone approached with respect, curiosity, or hostility. Building that reputation was a delicate art, a blend of showing enough toughness to deter attack but not so much to provoke a war you couldn't afford.

Resourcefulness became my refuge, a shield crafted from years of watching, waiting, and learning in silence. Food wasn't merely sustenance; it was a scavenger's prize, demands met with cunning and nerve. Black dumpsters behind shuttered convenience stores were treasure troves of discarded goods — stale bread, bruised fruit, even unopened packets of fast food tossed away without a second thought. But diving into those dark pits was never without risk; rival scavengers often lurked nearby, eyes gleaming with territorial claims and desperation. I learned to time my raids in the gray hours before dawn, slipping quietly through alleys while most were still cloaked in sleep, the only sounds my breathing and the steady thud of my heartbeat pounding in my ears. Water, too, was scarce and precious, often stolen from public fountains or bars, its scarcity driving many to make desperate choices.

Each day added new scars, both seen and invisible. The street was brutal in its lessons, breaking down those who didn't rise again with hardened skin and sharpened senses. Cold nights were torture, the biting wind a constant reminder of fragility. Sleep came in uneasy bursts, eyes flicking open to phantom footsteps or the shadows of figures moving just beyond the circle of weak light cast by a flickering streetlamp. The threat of violence was omnipresent—gangs marking territory with graffiti that read like ancient warnings, eyes that flashed knives in the dark, fists that could shatter bones in seconds. Fights erupted without notice, often over minor slights or scraps of stolen goods, the aftermath a tangle of sirens and red flashing lights piercing the darkness.

I was no stranger to such brawls, my body etched with reminders of each battle fought and won by sheer will. But survival was more than brute strength—it was knowing when to fight, when to flee, and when to disappear altogether. There was a fine line between courage and foolhardiness, and each misstep meant risking life and limb. Every scuffle was a narrative of desperation, a snapshot of the lives grinding against each other in this cruel ecosystem where the strong preyed on the weak, but even the strong lived with the constant fear of being overthrown.

The isolation was suffocating despite the crowds. Loneliness wrapped itself around me like a second skin, a constant companion more persistent than any shadow in the night. The world beyond the streets felt like a distant memory, a sepia-toned photograph slipping further out of focus with each passing day. While faces blurred into the same tired expressions of exhaustion and hopelessness, my mind became a fortress of memories—fragments of better times, the echo of laughter, the warmth of a mother's embrace—all swallowed by the oppressive present. This solitude fueled a restless hunger, driving me through endless nights of searching for purpose, or just a moment's peace amid the chaos. Yet even in that solitude, the street whispered its savage lessons—only the ruthless survived, only those who molded themselves into shadows could endure.

The art of evasion became second nature. Police cruisers glided like ghosts, their spotlights slicing through alleys and empty streets, each beam a threat to be avoided. To be caught meant cells smelling of decay and fear, locked spaces where hope shriveled under cold fluorescent lights. I learned to melt into any crack in the urban landscape—empty doorways, rusted fire escapes, beneath abandoned cars—always watching, always listening. Paranoia was a survival tool, a razor-edge between being caught or slipping away unnoticed. False friends, informants, and street justice created a complex web I could neither untangle nor trust. Even among supposed allies, whispers of betrayal lurked, a single wrong word condemning you to fade into the overlooked shadows or worse.

Bartering and hustling became the rhythm of my days. Skills developed on the fly—picking pockets with deft fingers, pressing a scam with practiced ease, or flipping small stolen goods for whatever scraps of currency could be scraped together. Trade was the lifeline on these unforgiving streets, where trust didn't exist, but transactions were a language everyone understood. The cycle of exploitation and scheming was endless; sometimes I played the victim, other times the predator. The street was a stage on which every role could flip on a dime, and survival meant mastering the script without losing what little soul remained.

Sleep deprivation gnawed at me, the nights haunted by nightmares of past violence and the ghosts of missed opportunities. The body ached, muscles twisted from sleeping on concrete, the constant high alert making rest nearly impossible. Still, the mind refused to quiet, replaying the decisions made and the ones avoided, pondering the impossible choices forced upon those with no stakes left but the fight itself. Hope flickered intermittently, like distant headlights cutting through a foggy night, but was quickly snuffed out by fresh hardships.

In this crucible, moments of human kindness shone like rare jewels amid the grime. A shared cigarette, a meal passed quietly between strangers, words of encouragement whispered beneath breath—all fragile threads connecting isolated souls stranded in the same storm. These fleeting connections rekindled the faintest glimmers of faith in something beyond survival. Yet, the instinct to protect oneself always nagged, a silent warning not to open too wide, not to trust too deeply lest the fragile bond be shattered. Love and loyalty here were precious and perilous, commodities traded cautiously.

The mental toll was relentless, weighing heavily as isolation bred paranoia and despair. Days blended into nights, memories sharpened by hardship circling like vultures in my head, torturing with reminders of past mistakes and lost chances. Yet, alongside the bleakness grew an iron resolve. The street taught me about limits—how far one could fall, how hard the ground would hit, and how necessary it was to push back. I learned to devour pain, turning it into fuel; wounds became badges of survival rather than surrender. My story was no longer just about living but about staking a claim in a world that had tried to write me off before I had even started.

Mirrors were scarce on the street, but I often glimpsed my reflection in shattered glass or city puddles. The eyes staring back were tired but unbroken, the set jaw firm against the torrent. That reflection told a story of someone fighting not just to exist but to find meaning beyond the grime and suffering. Every scar mapped the paths taken, each encounter etched lessons about humanity and its contradictions—the cruelty and compassion tangled inextricably in the same breath. The street was a prison, but also a strange crucible where transformation was carved out of hardship.

As days grew into weeks, I developed a rhythm in the chaos, a precarious balance that only those entrenched in street life could understand. Hunger, fear, exhaustion—they were constants, but survival hinged on more than enduring hardship; it demanded constant vigilance, unflinching adaptability, and the will to face down terror and loneliness alike. The streets had stripped me bare, exposing every weakness and vulnerability, but in that rawness, I found a flicker of strength. A stubborn ember refusing to be crushed by the weight of relentless darkness.

Hope, often mocked by the harsh realities, was nevertheless my secret weapon. It wasn't naive or pure—it was gritty and battered, forged in the fires of desperation and defiance. Survival was more than just avoiding death; it was about clinging to that spark, nursing it through endless nights stained with fear and loss until it cast enough light to reveal a path forward. Each day survived was a victory, a testament to a spirit refusing to be erased by the bleakness surrounding it.

That spirit was my armor, my currency, and my salvation. I became both the predator and guardian of my own existence, crafting a life from the fragments thrown my way. Street survival was the harshest education imaginable, but it was also a crucible that refined my senses, honed my instincts, and forged a resilience that carried me through every low point. In the shadows where others saw only despair, I glimpsed the faint outlines of redemption—a road jagged and treacherous but mine to claim if only I dared to keep moving forward.

Flickers of Humanity

The city had a heartbeat all its own, a rhythmic pulse fueled by the relentless hum of sirens, shouted arguments, and the distant clang of steel on concrete. But beneath the cacophony and chaos, there were moments where the relentless grind of street life softened just enough to reveal fleeting glimpses of something profoundly human— tenderness, understanding, hope—even if those moments were brief and often cloaked in the grit and grime of survival. Flickers of humanity shone through the cracks and craters of my world, little acts and encounters that reminded me I wasn't entirely invisible, despite the sinking feeling that the shadows were swallowing me whole.

On the coldest nights, when the wind bit through my threadbare jacket and the roar of the city seemed intent on tearing my spirit apart, I found refuge not in some grand sanctuary but in the simplest gestures exchanged with those who knew the street as intimately as I did. There was Old Joe, a man with a face like weathered leather and eyes deep with stories no one asked to hear. He never had much to give, but one night he pressed a small, lukewarm coffee cup into my hands, his trembling fingers brushing mine briefly. "Keep your head, kid," he muttered, voice cracked by years of smoke and regret. That cup, barely enough to stave off the chill, felt like a lifeline tethering me to some shred of kindness, an unspoken acknowledgment that I was seen—not just as another nameless shadow, but as a person wrestling with pain and desperation.

The streets were a jungle of distrust and sharp edges, a place where kindness could get you stabbed in the back as easily as it could forge a fragile bond. I learned quickly to weigh every smile, every outstretched hand, measuring sincerity versus self-interest. It wasn't easy to let anyone close, to risk the vulnerability that love or empathy demanded when betrayal lurked in every alleyway. But sometimes, between the flashes of gunfire and the rush of adrenaline that addiction and violence pumped into my veins, there were pauses—breathless spaces where conversations stretched longer than survival allowed and hands were extended in genuine concern. I remember Maya, a woman draped in a mishmash of faded colors and a worn scarf that once might have been bright, who shared her meager sandwich with me on one scalding afternoon. Her eyes, heavy with exhaustion but alight with fierce kindness, never asked why I'd been hungry or how I'd come to stand on the edge of the street begging for scraps. She just saw me, and in her silent acceptance, I found a rare moment of peace.

Isolation had been my closest companion for as long as I could remember, a cold, empty darkness that whispered lies about my worth and left me gasping for something real amid the rubble. Yet, paradoxically, it was in those moments of shared hardship that I began to glimpse the frailty and strength of human connection. The woman who called herself "Rose" sat with me once beneath the flickering light of a broken streetlamp, her leg wrapped in a stained bandage as the city around us roared in relentless motion. She was a junkie like me, her hands shaking, her voice fragile but steady. We talked softly about lost hopes and broken dreams, our admissions wrapped in the kind of honesty street life rarely tolerated. It was one of the few times I dared to let myself be vulnerable, to admit that beneath the hardened exterior there was a boy desperate to be known, to be saved. Rose's smiles came and went like dawn, quick and elusive, but when they did, they illuminated the dark corners of my soul.

Not all encounters carried warmth, of course. Some were brutal lessons disguised as kindness. There was the moment when I accepted shelter from a man who claimed to be a worker for a local outreach, only to discover later that his generosity came with demands— promises exchanged for control and pain. But even those betrayals carved lessons into my memory, teaching me discernment, teaching me that flickers of humanity could sometimes be smokescreens hiding something far darker. Still, each cruel encounter only sharpened my will to seek those rare luminous connections, to find the threads of compassion that wove through the tapestry of misery.

In the midst of all this, my resourcefulness became my lifeline. The streets demanded it. I learned to read people like the pages of a book—every twitch, every glance, every hesitation coded into a language of survival. It was this intuition, this sharpened awareness, born from countless nights weaving through shadows, that guided me to moments of unexpected grace. Like the young man who once slipped me a warm slice of pizza from a corner shop without saying a word, his eyes flickering with the ghost of a smile. It was a simple act, but in that gesture, I felt a flicker of brotherhood that transcended the grime and desperation.

One time, I found myself sitting next to a runaway girl no older than fifteen, huddled beneath a thin blanket that barely kept out the biting cold. She was trembling, eyes darting toward the alley's mouth every few seconds, terrified of the ghosts chasing her. Without a word, I took off the worn jacket falling from my shoulders and draped it over hers, our bodies blending under its embrace like two fragments of broken glass fitting together. She didn't say thank you, but her quiet sobs softened, and I sensed an unspoken trust beginning to take root in that tiny shared space. Moments like that complicated the narrative I'd carried about those on the fringes of society—the assumed coldness, the hardened hearts. They were alive with fear and hope in equal parts, just like me.

Throughout the spiral of my addiction and the chaos of survival, I encountered acts of kindness that refused to be erased by the darkness. The volunteer at the local soup kitchen who, despite my scowl and sullen silence, insisted on saving me an extra cup of coffee every time I came through the line; the old woman who handed me a folded dollar bill when no one was watching, whispering, "Keep fighting; you're worth more than this;" the way a stray dog briefly accepted me into its wary pack, a reminder that trust could exist, even if fleeting. These encounters stood like beacons, illuminating parts of the city and myself that still harbored hope—tiny embers beneath the ash of my despair.

Yet, even as these flickers of humanity flickered and faded, I realized the harsh truth that kindness often came swaddled in pain, that every act of grace on the streets was both a relief and a reminder of the immense suffering that made it so precious. The same gestures that brought warmth also underscored the brutal conditions forcing people into survival mode. No one offered a warm cup or a stolen sandwich without carrying their own burdens, their own scars, and their own stories saturated with loss. Marks of addiction, abandonment, and violence were etched into every face I passed, each person a reflection of collective pain and resilience.

What struck me most profoundly was not just the acts of kindness themselves, but the courage it took to offer them amid chaos. To extend a hand amid gunfire required a faith in humanity that felt radical in a world designed to break spirits. Each moment of compassion became a thread in the fraying fabric of a city struggling to hold on to its soul. And for me, each thread was a reminder that redemption wasn't a solitary journey but one that depended on others—on those who dared to see the brokenness and respond not with judgment, but with empathy.

I remember an evening when the rain pelted the pavement like a thousand pinpricks, turning the city into a blur of shadows and light. I'd taken shelter in the doorway of a shuttered storefront, exhausted and drenched, my body protesting the cold that crept into my bones. A couple passed, seemingly oblivious to the storm or to me. But then the woman dropped her umbrella with a soft clatter near my feet and stopped. Instead of walking on, she knelt down, scooped it up, and with a look that folded pity and quiet strength into one gesture, she gave me the umbrella's handle. It was the kind of simple, unspoken kindness that flooded the small corners of my soul with warmth despite the damp and despair. I gripped that umbrella like a talisman, something more precious than any drug or dollar, a symbol that beneath the ugliness, beneath the grit, there was still light worth chasing.

Occasionally, even the hardened gang members who ruled the streets with iron fists revealed unexpected humanity. There was Tyrone, whose reputation for violence preceded him like a dark cloud, yet whose acts of mercy saved my skin more than once. He once risked his standing by stepping between a younger kid and a rival gang, his voice low and gravelly as he defused a violent standoff. Later, in a rare moment away from the reign of intimidation, he shared a smoke and a laugh with me, revealing cracks in his armor and stories of a childhood lost to grief and neglect. His presence complicated my black-and-white view of friends and enemies, reminding me that beneath the surface of violence were fractured souls aching for connection.

The moments of kindness I stole from the unforgiving streets were sometimes so brief and delicate that I feared blowing on them too hard, lest they vanish. They often came without warning, like a beacon in a storm, a contrast to the raw despair that surrounded us all. Those flickers of humanity became the glue holding together the shards of my battered self, a fragile yet vital proof that even in the depths of addiction and violence, the capacity for care endured.

Even in my darkest hours, when the weight of addiction pressed on me like an anchor dragging me deeper into the abyss, I found that kindness could serve as a light—sometimes flickering, sometimes faint, but enough to guide me back. Whether it was the compassionate nurse who never gave up on me despite my anger and self-destruction, or the counselors who listened without judgment during therapy sessions that cracked open my defenses, these encounters slowly chipped away at the fortress of isolation I'd built. They infused bitter survival with glimmers of hope, underscoring the truth that healing was not just about ceasing the use or escaping the past, but about rebuilding trust in the very fabric of human connection.

Resourcefulness lay at the heart of navigating this complex landscape of hardship and fleeting grace. To survive, I needed to cultivate not just the instinct of self-preservation but also the ability to recognize and nurture these sparks of kindness in others and in myself. I learned to carry the memories of those moments like a secret stash of warmth, drawing on them when the world closed in, and my resolve faltered. They became the narrative threads that rewrote the story I told myself—a story not solely of darkness and despair, but one where light could pierce through, persistent and real.

These flickers of humanity did not erase the brutality of my existence, nor did they absolve the pain that addiction and violence had wrought. Instead, they formed a delicate balance, the yin to the yang of my experience. Each act of kindness challenged the harshness, and each moment of empathy carved out a space where healing could begin. It was these moments that seeded my eventual belief in redemption, teaching me that no matter how deep the shadows stretched, the human heart could still find its way toward the light.

Looking back, I realize that these encounters—small and seemingly insignificant to the indifferent eye—were the anchors of my journey. They reminded me that I was more than my mistakes, more than my addiction, more than the violence that threatened to swallow me whole. They were proofs that even in the most broken places, the potential for love and humanity remained, waiting to be found, waiting to be embraced. It is in these flickers, however brief and fragile, that I learned the true meaning of resilience—not just surviving, but finding the courage to hope, to connect, and ultimately, to reclaim a life beyond the shadows.

Chapter 8
Therapy and Truth

Seeking Help

I remember the exact moment when the idea of seeking help shifted from a blur in the back of my mind to a stubborn, unshakable imperative. It wasn't wrapped in some grand revelation or miraculous event; no lightning bolt from the heavens came crashing down on me. Instead, it was a quiet accumulation of weariness—a crushing exhaustion from running on fumes, from begging for peace in a storm that refused to abate. You carry your survival stories like scars etched deep within your skin, and for the longest time, I carried mine as if they were badges of honor. That's what I told myself, anyway. Each scar was a testament to how much I'd endured, how far I'd fallen, and still managed to claw my way back. But underneath that defiant pride was a raw, exposed nerve—a part of me that begged for reprieve, even as I resisted with every fiber of my being because asking for help meant admitting just how broken I was.

The decision to enter therapy wasn't an easy one. It felt like stepping off a cliff into unknown territory with no promise of a net. There was this gnawing voice inside me, a cacophony of doubts and fears screaming louder than any encouragement. What if I wasn't "broken" enough to warrant help? What if I was? More than that, what if admitting I needed therapy made me weaker? In the streets, showing weakness was like inviting vultures to pick at your bones. The tough exterior was armor hardened through years of battling both external enemies and internal demons. But that armor came with a price—a suffocating loneliness, an impossible weight to carry. To make the leap, I had to face the sprawling shadows of my past—everything I'd buried deep beneath layers of denial and bravado.

The first therapy session tasted bitter on my tongue, like swallowing a harsh truth I wasn't ready to digest. The room was sterile but strangely intimate, a small, sunlit space that seemed worlds away from the alleyways and sirens that had been the soundtrack of my life. The chair felt alien beneath me, the silence heavier than any street confrontation I'd encountered. Across from me sat a woman whose calm eyes held no judgment, just an invitation. I was hesitant, words catching in my throat as I tried to explain the unexplainable chaos I carried—fractured memories, tangled feelings, trauma wrapped up in layers of addiction and violence. The sessions felt like staring into a cracked mirror; every reflection fragmented, forcing me to confront parts of myself I didn't want to see. But there was also something quietly hopeful in that pain, a trembling awakening to the possibility that maybe, just maybe, I could be more than the sum of my broken pieces.

Resistance came in waves. Some days, I wanted to storm out of that room and slam the door shut on everything I was told to unearth. There were moments when nostalgia for the streets' brutal clarity beckoned me back—the rawness of survival, the numbness I could rely on. Therapy was uncomfortable, raw, and at times, it felt like I was bleeding in front of a stranger. There were tears I thought I'd run out of—old griefs that resurfaced like ghosts I'd tried to exorcise. But amidst the discomfort, there were small breakthroughs—snippets of language that began to sculpt new frameworks for understanding myself and my story. It was in those moments that I started to realize resilience wasn't about stoic silence or enduring pain alone. It was about letting that pain speak, and daring to listen.

The path to healing unfolded unevenly, like the jagged rhythm of my own breath after a sprint through dark streets. Some days brought clarity that felt like the dawn breaking through a long night; others were storm clouds threatening to swallow me in doubt and despair. The therapist's questions were gentle prods but sometimes felt like pebbles thrown into a still pond, stirring up ripples I wasn't prepared to face: Why did I start down this path? What wounds did I carry that drove me towards danger and addiction? Why did I cling to the people and places that threatened to tear me apart? Each answer was a confrontation with vulnerability that was both terrifying and liberating. It exposed the worn seams of a heart hardened by circumstance, yet beneath that grit, I found enduring threads of hope.

There was a recurring challenge in disentangling who I was from what I'd done. In the streets, your actions define your identity—gang affiliations, survival strategies, the ghosts of addiction. But therapy forced me to see beyond the surface narrative, to confront the boy who had been desperate to survive but forgotten to live. It meant owning my mistakes without letting them imprison me, recognizing the humanity in my darkest moments. That kind of reckoning didn't come easy. Some days, it felt like I was peeling off my own skin, raw and vulnerable to the world's harsh gaze. But with each session, the idea of healing shifted from a distant luxury into a tangible possibility.

I began to unpack the relentless guilt that had anchored me during those years. The guilt for the people I'd hurt, the chances I'd squandered, and the innocence that addiction and violence stole from me. It was a weight that screamed in the silence, gnawing at my conscience and sapping my will. Therapy didn't magically erase those pangs but helped me understand that guilt was a signal, not a sentence. It was part of the journey, a reminder of where I came from but not a verdict on where I could go. I learned to meet that guilt with compassion, to let it inform my growth rather than suffocate it.

Trust was the hardest currency to earn in those sessions. Opening up meant gambling with vulnerability in a world where survival had been about closing off, building walls that kept pain at bay. The therapist became a mirror reflecting my fractured truths back at me without distortion or shame. Over time, the sessions felt less like interviews or interrogations and more like conversations with a guide leading me out of a labyrinth. I realized that recovery wasn't a linear climb but a twisted path with setbacks and leaps forward. Every moment of honesty was a quiet victory against the chaos that had once seemed relentless.

One of the most profound realizations came when I started to recognize the patterns that tethered me to self-destruction. The addiction wasn't just chemical—it was emotional and psychological—a desperate attempt to silence the noises inside that no one else could hear. Therapy helped me map those internal landscapes: the fears, the hurts, the abandoned hopes. It was like learning a foreign language that my own mind had spoken in for years but refused to translate. Slowly, the sessions began to offer me tools—not just for coping but for transformation. I learned to confront cravings not with shame but with awareness, to challenge the old narratives that said I was destined to fall again and again.

There were breakthrough conversations that unlocked doors I had slammed shut for years. One session, I found myself telling the therapist about the night I was nearly killed, the moment when survival felt like a fractured, fragile thread slipping through my fingers. Speaking those words aloud gave them weight, and the weight made them real. For the first time, the trauma wasn't just a shadow following me; it was a story I could own. In that owning was power—the power to reshape meaning, reclaim agency, and envision a life not ruled by fear but by possibility.

Even as I bared my soul in those rooms, part of me was still on guard. Shadows of doubt lurked at the edges—what if I got too vulnerable? What if the world outside therapy's safe walls would judge and reject me? But in time, I learned that healing required taking those risks. It meant creating spaces where brokenness was met with empathy, not condemnation. Therapy became a sanctuary, a place where my scars transformed from chains into maps leading to wholeness.

The emotional rollercoaster inside those sessions reflected the chaos of my past but also modeled a new rhythm of living—a rhythm where pain was acknowledged but not allowed to dictate the future. There were days when the tears spilled freely, a cleansing torrent that washed away years of accumulated despair. Other days, silence spoke volumes—the weight of unspoken fears settling quietly between words. Yet even in the silence, there was a stubborn pulse of hope, a belief that the messiness of healing was worth the struggle.

Therapy also revealed how deeply interconnected my past traumas were with my sense of identity and belonging. My addiction and gang involvement had been attempts to carve out a place in a world that continually told me I was invisible or disposable. Confronting that truth was humbling and painful—it meant rewiring my understanding of self-worth away from external validation and toward internal acceptance. It was a journey of unlearning the harmful scripts that had guided my actions and replacing them with new narratives of resilience and love.

That new narrative didn't come without moments of doubt and despair. There were times I questioned whether change was possible or if I was condemned to repeat old patterns. The sessions often surfaced uncomfortable memories—the betrayals, the losses, the nights of numb desperation—that felt like they would drown me if I let them. But the therapist's unwavering presence provided a lighthouse amidst those storms, a steady anchor that encouraged me to keep sailing even when the waters were rough.

Over time, therapy became less about recounting traumatic events and more about actively shaping my future. I began setting goals—not just to stay clean, but to understand who I was beneath the layers of survival instinct and addiction. I dared to imagine a life defined not by chaos but by purpose. With each breakthrough, no matter how small, the fog of my past lifted slightly, revealing glimpses of clarity and peace I hadn't known was possible.

Therapy also coaxed open spaces in my soul for forgiveness—both for others and for myself. I started to release the bitterness that had calcified inside me, realizing that clinging to pain only kept me tethered to the past. Forgiveness wasn't about absolving wrongs but freeing my own heart to breathe without the weight of resentment. This shift was subtle but profound—it began to chip away at the barriers that had kept me isolated for so long.

The emotional turbulence of therapy often spilled into my everyday life in unexpected ways. Relationships that once seemed irreparably damaged became arenas for practicing new communication built on honesty and vulnerability. The courage it took to face my own demons gave me the strength to reach out and rebuild connections. Though fragile and imperfect, these attempts at repair filled the cracks inside me with light and possibility.

Despite the progress, the journey was not linear or predictable. There were setbacks—the nights when cravings felt unbearable, when memories threatened to overwhelm, and moments when old coping mechanisms beckoned like sirens in the dark. But therapy had given me an arsenal beyond just willpower; it provided understanding, tools, and a community of support that buffered against the pull of relapse.

In the quiet moments between sessions, I reflected on how therapy was reshaping my relationship to pain itself. Rather than avoiding or numbing it, I was learning to sit with discomfort—

allowing it to teach me rather than break me. This shift was revolutionary, inviting me into a new way of being that honored both struggle and strength. It was a dance between acceptance and action, a constant negotiation between darkness and light.

Looking back, seeking help was the bravest, most vulnerable act I undertook. It meant dismantling the barriers I'd built against the world, exposing wounds that had festered in silence, and daring to believe that healing was not only possible but within reach. The path was jagged and steep, marked by fierce battles with fear and shame. Yet in choosing therapy, I chose myself—flawed, broken, yes, but also capable of growth, worthy of care, and destined for redemption beyond the shadows that once defined me.

Each session was a step out of the shadows, a reclamation of the parts of my soul that had been lost to violence and addiction. It was a testament to the resilience that had carried me through the darkest nights and now illuminated a future shaped not by despair but by hope. Seeking help was not a sign of weakness but a declaration of survival, a commitment to the hard, beautiful work of becoming whole again.

Facing the Past

In the dimly lit room where I first stepped for therapy, the air was thick with a silence that stretched between every word unspoken, a quiet so heavy it almost bore down on my chest like the weight of years I hadn't dared to unpack. The walls, a muted gray, seemed to press inward, conspiring with my own defenses to keep everything locked away—memories buried beneath layers of pain, shame, and survival tactics honed on the unforgiving streets and hardened by years of addiction and violence. Facing the past was never going to be a neat, linear process; it was a brutal excavation of everything I had spent decades trying to forget or bury under numbness. With each session, I found myself peeling back the protective layers I had built, revealing not only the scars etched by trauma but the raw, bleeding wounds beneath that still ached like they had been ripped open mere moments ago. As the therapist's voice gently nudged me toward those buried memories, I wrestled with the sudden rush of feelings I had suppressed—fear, anger, grief, and an aching loneliness that threatened to swallow me whole.

What surprised me most was how the past didn't come neatly packaged, ready to be confronted in a tidy sequence. Instead, memories slammed into me sporadically—flashes of my mother's worried eyes, the gunfire echoing in my neighborhood, moments of betrayal by those I once called brothers in the gang, and that cold, numbing hunger for whatever momentary escape drugs could provide. These fragments collided inside me like shards of glass, cutting deeper with every recall. Sometimes, I would leave the sessions breathless, heart pounding in a way that mimicked the adrenaline highs of my darkest days. Yet, in those moments of vulnerability, there was a small flicker of something else—hope. Therapy wasn't just a relentless digging into pain; it was, slowly, the light filtering through the cracks of years spent in darkness.

The first breakthrough came inattentively, almost as if the moment weren't mine to control. We were sitting across from each other one chilly afternoon, the thin frost outside reminding me of the cold nights spent shivering under the bridges, the nights when I first learned to dread silence because it was often broken by screams. My therapist gently asked about my earliest memories, but instead of the usual detachment, something shifted. I found myself transported back to my childhood home—not the house, but the feelings. The warmth of my grandmother's kitchen, the way her hands would cradle a cup of chipped coffee mug, the smell of cinnamon and old wood mingling with her soft humming as she sang church hymns. Those sounds, smells, and moments seemed incongruent with the pain I expected to revisit, but they were there, nestled beneath the chaos. In that glimpse of tenderness, I realized how much had been lost in all the chaos and violence that consumed my adolescence. That day, I cried, not just for the pain, but for what had once been and might yet be again.

But healing was not a steady climb. There were days when the floodgates opened so wide I felt like I was drowning. One session unfurled the storm of guilt that had wrapped around me for years like barbed wire. It was after I opened up about the near-fatal shooting I survived, and the lives lost in the street battles that followed. The faces of everyone I had hurt, knowingly or unknowingly, came crashing down in my mind—the kid I'd pushed too far in a petty rage, the woman caught in the crossfire, my own brother who turned his back on me because he could no longer bear the life I was leading. The weight of these memories became so overwhelming that I could barely speak; my voice cracked, and tears blurred my vision. I was forced to confront the harsher truths of my past—the ways I had been both victim and perpetrator, the tangled morality of survival in a world that offered no easy choices. And yet, in the midst of that pain, my therapist's unwavering support gave me the courage to hold myself with a sliver of compassion, to see myself not just as a villain in my own story but as a broken man trying to piece together what was left.

The journey into those shadows revealed the complexity of trauma—the way it distorts time and memory, how events blur and overlap in a chaotic mosaic that left me grasping for understanding. I began to realize that my addiction had been more than a series of bad choices; it had been a desperate attempt to silence the demons screaming inside me, to anesthetize memories that no child should carry. The small victories came in moments like learning to identify the triggers—certain smells, sounds, or even weather patterns—that would send me spiraling back into the past. For example, the sound of police sirens was once enough to catapult me into a panic attack, back to the sting of handcuffs and the cold concrete of a jail cell. Therapy sessions became a space to practice gentler responses, to rewrite the conditioned reactions that had governed my survival.

One of the most daunting challenges was untangling the distorted relationship I had with love and trust. In recounting my failed relationships, especially my brief and turbulent connection with the Canadian escort, I saw clearly how much fear had influenced my actions. The walls I built around my heart, designed to keep pain at bay, had also shut out healing. I was terrified of vulnerability, of allowing someone the power to see my brokenness, of risking more abandonment. Yet, the paradox was that it was through that very relationship—flawed, messy, and complicated—that I first began to glimpse the possibility of intimacy beyond survival. Revisiting those memories brought fresh waves of grief but also gratitude for opening me to the idea that love need not always be a weapon or a wound.

Therapy was not just reliving the past but reconstructing a narrative where I was no longer defined by my darkest moments. The repetitive patterns of self-defeating thoughts and despair that once swept over me began to loosen their grip. Slowly, I nurtured seeds of self-forgiveness and resilience that had lain dormant beneath years of self-loathing. There were exercises that stressed mindfulness, teaching me to anchor myself in the present when memories threatened to overwhelm. Sometimes, I would sit in silent meditation, tracing the path from despair to hope like a fragile thread woven between the shards of my past experiences.

In facing the past, I also confronted the elusive sense of identity. Who was I beyond the labels of addict, gang member, convict, and survivor? Each session nudged me toward pieces of myself I had lost touch with—the hopeful child who dreamed of a better life, the young man who yearned for respect but not at the cost of his soul, the person capable of kindness beneath layers of defense. This identity reconstruction wasn't comfortable; it often rattled the foundations I had relied upon for so long. But it was necessary work to reclaim a self untethered from trauma's defining grip.

The narrative from these therapy sessions was punctuated by moments of rage—not only at the injustices I had faced externally but at myself for the choices that led to destruction. Expressing this anger was freeing but also terrifying, unearthing feelings of powerlessness I had long ignored. There were tears shed for lost time, friendships, and family ties broken by addiction and violence. The path to healing wove through forgiveness—not only seeking it from those I had wronged but also granting it to myself, an act far harder than I'd imagined.

Amidst this tumult, I learned that healing was not a solitary endeavor. The support woven by groups and advocates I later engaged with became a lifeline. Sharing pieces of my story, once a source of shame, transformed into a beacon of connection and hope—not just for me, but for others wrestling with similar battles. The courage to face the past shifted into a commitment to social change, driven by the hard-won empathy born from surviving what once seemed insurmountable.

Each session drew me deeper into that fragile space between past and present, where pain coexisted with possibility. I often left feeling raw yet renewed, every painful recollection a step toward reclaiming the narrative of my life. And though the journey was far from complete, the act of facing the past became a testament to resilience— a refusal to be chained by history, and instead to wield it as a source of strength and understanding. The memories no longer felt like weapons aimed at my heart but pieces of a mosaic that, when viewed whole, told a story not merely of survival, but of transformation.

Steps Forward

The first glimmers of light after such deep and relentless darkness have a tenuous quality, like fragile threads of dawn weaving timidly through the collapsing blackness. It was in this tenuous space, filled with both the surrender of old defenses and the sudden stirrings of emerging hope, that my journey through therapy truly began to find its footing. Each session was a battleground and a sanctuary, a place where the armor I'd so meticulously crafted over years of survival was both challenged and slowly dismantled. The walls I'd built around my heart and mind, forged from pain, distrust, and hardened necessity, gradually became permeable, allowing the first drops of vulnerability to slip through. It was terrifying and liberating in equal measure — to confront the shadows I'd long buried, especially when those shadows were sharp and jagged with raw memories of violence, loss, and addiction.

Those initial sessions were marked by a peculiar kind of silence — the type filled with uneasy tension. I would sit across from the therapist's calm eyes, the room sterile yet strangely intimate, feeling parts of myself resist like a wild animal trapped behind invisible bars. There was a braided rhythm to these encounters; my words often came out in halting bursts or through reluctant admissions. The deepest struggles were not always verbalized, clutched too tightly in the places where shame and fear converged. But beneath the surface resistance, the most essential work was unfolding. The therapist's voice was steady and patient, never pushing too hard, but persistent enough to chip away at the fortress of my avoidance. Sometimes, it was a simple question — a prompt that forced me to confront feelings I'd learned to suppress decades ago: What did I feel when I saw my friends fall to the streets? How did I survive the silence of my own mind screaming through withdrawals? What did love even mean when it was so intertwined with betrayal and loss?

Each answer I gave, spoken slowly and often swallowed by tears or gritted jaws, was a step forward — raw and imperfect but real. I learned to recognize that healing was not linear; it was fractured, messy, with setbacks folded into the advances. There were days when progress felt like a distant promise; I left sessions haunted by the intensity of what I had dredged up, my whole body aching with the weight of memories that were sharper than any wound. I recall one particular session where recounting the night of the shooting left me trembling; the sounds replayed in my head with excruciating clarity — the piercing bang, the shocked silence, the cold blood that ran beneath the cracked concrete like a sinister river. Yet, instead of retreating inward, I found something unexpected stirring: a flicker of courage to claim that story, to hold it without being consumed by it. That's what therapy began to teach me — how to sit with painful truths, how to let grief exist without being paralyzed by it.

The emotional challenges were relentless. Tremors of guilt for the people I left behind, swelling waves of anger that I initially directed at myself, and sometimes at the systems that had failed us all, bubbled up in unpredictable ways. These sessions opened cracks for introspection I hadn't dared enter before. The vulnerability was exhausting; opening old wounds meant exposing soft places to the brutal light of self-examination. But gradually the sessions built a vocabulary for experiences that had long been nameless: childhood trauma, complex grief, post-traumatic stress, addiction's grip. Naming these truths granted me a semblance of control, an avenue out of the chaos. The therapist's tools — grounding exercises, mindfulness techniques, cognitive reframing — became lifelines in moments when my mind threatened to spiral into despair or rage. I remember sitting in the quiet of the room, learning slower, deliberate breathing, feeling the pulse of life beneath the skin of my trauma. Breath by breath, the sharp edges dulled.

One of the biggest breakthroughs wasn't simply from talking but from the way the sessions illuminated patterns I hadn't discerned before. The cycles of violence, self-sabotage, and isolation fell into relief, and suddenly I saw how each decision was tied to survival instincts deeply embedded in the streets I'd known too well. This realization didn't excuse the damage I caused or the betrayals I endured, but it gave context and, crucially, a pathway toward change. I began to understand that my identity was not fixed in the streets, the gangs, or the addiction — rather, there was a self underneath all the scars, bruises, and failures that could be nurtured and grown. Therapy cracked open a new narrative where hope and despair coexisted, a story I might rewrite with the tools I was gaining.

At times, the progress was subtle — a shift in how I regarded myself after a session, or an internal dialogue that replaced automatic self-condemnation with tentative kindness. At other moments, it was seismic: the first time I expressed anger without fear, or the day I allowed tears to flow freely without dragging shame in their wake. These breakthroughs weren't grand gestures but small, profound acts of reclaiming my humanity. I remember how powerful it felt just to say, "I am worth healing," a phrase I had never said out loud to anyone, least of all to myself. The daily monotony of survival had been replaced by a flickering awareness that life could hold more — more meaning, more love, more peace.

Friends and family noticed changes too, subtle shifts in my demeanor that hinted at the struggle and hope simmering beneath the surface. There was skepticism, of course, from those who'd seen me fall before, who expected relapse or defeat. I wasn't naïve — I knew the shadows would always lurk, that the temptations to numb pain or escape reality wouldn't vanish overnight. But each day that I made a conscious choice not to sink back into old patterns was a victory, a foothold on a steep and rocky ascent. The cracks of my past no longer felt like abysses but like fissures, spaces where light could eventually seep through.

Sessions also unearthed new layers of pain surrounding my relationships — fractured bonds with my family, the complex entanglement with the Canadian escort, and the ghosts of friends lost to violence and addiction. I realized that healing myself depended on untangling these webs with honesty and compassion. Forgiveness — both of others and myself — was a concept I wrestled with fiercely, often overwhelmed by the enormity of past betrayals and failures. The therapist guided me gently into this terrain, reminding me that forgiveness wasn't about condoning harm but about freeing myself from the chains of bitterness that kept me tethered to my torment. It was a slow process, filled with resistance and retreat, but as I moved deeper into it, a sense of peace began to emerge.

Outside of therapy, the world felt harsher at times. The streets hadn't changed much, nor had some of the systemic inequalities that had shaped my life from the start. Moments of hopelessness would surface unexpectedly: a trigger in a song, a news story about another shooting, the sight of someone else caught in cycles I was desperate to leave behind. These moments tested the fragile hope I carried within, chipping away at my resolve. Yet, the foundation therapy provided allowed me to withstand these storms with a new kind of grit — a mixture of humility and defiance. I refused to let myself be swallowed again. The power of confession — of laying my rawest struggles before another human — had created an unexpected sense of solidarity, a reminder that I wasn't alone in this fight.

In the quiet hours between sessions, I spent long stretches reflecting on the work done, journaling my thoughts and emotions in illegible scribbles and messy pages. Writing became another way to channel the storm inside, to transform the chaos of memories into something tangible, something I could hold and examine, rather than letting the pain churn endlessly within. As I looked back at the ink-stained pages, I saw the evolution of my mind — from rage and desperation to moments of clarity and acceptance. These acts of self-expression fortified me, filling gaps left by isolation and trauma. I also began reaching out to support groups, hesitant at first, but soon recognizing the strength found in shared stories, the unspoken understanding that linked us beyond words.

What surprised me most about the gradual healing was how it changed not just my internal world but the way I started to navigate life externally. I began to set small goals: attending appointments regularly, maintaining a healthier routine, reconnecting with family members in cautious but sincere attempts at reconciliation. There was a tension between hope and fear every step of the way — the fear of relapse, the fear that the past would always define me — but hope kept burning as a quiet ember, fanned by each tiny success. I remember walking through my old neighborhood, feeling the ghosts of my former self whispering in every corner, yet this time I saw it not as a trap, but as a landscape I could traverse without being consumed. The streets were no longer the sole arbiter of my fate.

Therapy also opened doors to recognize and claim the parts of myself that had been silenced for so long: the creative sparks, the desire to help others, and the capacity for deep empathy. I started to imagine a future beyond survival — a future where my experiences could serve a purpose not just for me, but for others languishing in similar darkness. This nascent sense of purpose was fragile but vital. It gave me a reason to withstand difficult days, to endure the hard labor of self-confrontation and change. There was still a long way to go, but each session, each breakthrough, no matter how small, marked a step forward on a path that was no longer invisible or unreachable.

I learned that recovery was not about erasing the past or pretending it didn't hurt, but about creating new narratives where pain coexisted with hope, where wounds could heal without erasing scars. The courage to face my story, with all its brutal truths and moments of grace, was itself a radical act of defiance against the darkness that had once threatened to erase me. In those early steps forward, amidst setbacks and breakthroughs, I found that healing was not a destination, but a relentless, courageous process — an act of faith in the possibility of light after enduring so much shadow.

Chapter 9
Sober Reflections

New Routines

The mornings, once a blur of dread and desperation, began to take on an unfamiliar but welcome rhythm. It wasn't easy, not by any stretch of the imagination. Each day was a battlefield where the enemy hid not only in the world outside but also deep within my own mind. The gnawing cravings, the flickers of doubt, the unbearable weight of past mistakes—they all waited for me to falter, to let my guard down for even a moment. But I learned quickly that in this war for my life, the smallest victory was in showing up. Getting out of bed on time, stretching my stiff limbs, facing the mirror without flinching—all those acts of defiance against the pull of the old ways slowly built a new kind of foundation beneath me. I discovered that when I consciously shaped my mornings, setting intentions with deliberate care, the rest of the day often followed suit, however unpredictable it might remain.

At first, my mornings began with simple rituals: a glass of water to clear the taste of the night's turmoil, a few deep breaths punctuating the silence before the chaos of the city seeped through the cracked window panes. I forced myself to sit quietly, though the silence was deafening—a stark contrast to the constant noise of my former life. Those few minutes, undoubtedly the hardest, were where I wrestled with the ghosts that clawed their way forward just as I thought they might be laid to rest. Memories of streets laden with the brown haze of smoke, sirens wailing, the echo of shouted threats, and the hollow ache of addiction all crowded the small space in my head. It was in those solitary moments that I confronted the rawness of my trauma, sometimes with tears, sometimes with clenched fists against the unrelenting tide of pain. Each morning, I chose not to drown in the past. Each morning, I chose to breathe instead.

The routine grew, slowly stretching like the dawn light across a darkened room. I added journaling to the mix, scribbling thoughts and fears, no matter how fragmented or ugly they seemed. Pouring pen onto paper became a lifeline—an honest conversation with myself, devoid of the judgment that so often accompanied my internal struggles. Writing was not artistry; it was survival. By capturing the volatile storms inside me, I could begin to understand their shape and size, naming the fear and anger that had long remained buried beneath layers of denial. The words that fell from my pen were jagged and raw, like the streets I'd left behind, but in that rawness, there was a spark of something new—hope, perhaps, or at the very least, a sliver of clarity. Through writing, I started to see patterns in my thoughts, recognize triggers that set off the gnawing need for escape, and acknowledge the moments of strength I had denied myself for so long. It was the beginning of self-discovery, messy and halting, and deeply necessary.

But new routines could not take root in the quiet refuge of my room alone; they had to be tested in the unforgiving light of day. Venturing into the world sober was itself a challenge that demanded both courage and careful planning. The streets, with their familiar smells and sounds, still whispered temptations, promising relief and oblivion. I knew that denial or complacency could become my undoing. It was here, in those hours outside the walls of safety, that I leaned heavily on coping mechanisms—tools I had to forge from the furnace of my struggle. Exercise became one of those pillars. Long walks through neighborhoods scarred by the same epidemic I'd survived, the crunch of gravel beneath my feet, or bursts of jogging that pushed my lungs and legs to remember what it felt like to be strong in a new way—these physical acts were more than just attempts at health. They were an assertion of control, a reclaiming of my body from the addictions that had once owned it. The rhythm of movement synchronized with breath, gradually clearing the fog that addiction had cast over my senses.

Therapy also became a cornerstone of my new life. Attending sessions was sometimes a battle in itself—dragging my weary self through the door, bearing wounds that were still fresh, exposing vulnerabilities previously buried beneath layers of toughness and street-born pride. But each encounter peeled back a layer of shame and secrecy, making space for understanding and, eventually, forgiveness. My therapist became a witness to a raw truth I often refused to face, a mirror reflecting not just my mistakes but also my resilience. Through guided conversations, I learned to dismantle the self-destructive narratives I had internalized, replacing them with narratives of worthiness and possibility. Therapy was not a cure-all—it was a path strewn with jagged rocks—but it was vital in breaking the vicious cycle where despair often turned on itself. It was a place where hope could, slowly and tentatively, take root.

Support groups were another critical thread woven into the fabric of my new routine. Surrounded by others bearing their own scars and stories, I found an unexpected kind of family. There was something profoundly healing about the simple act of listening to another's confession and finding pieces of my own pain whispered back in their words. The shared spaces of vulnerability bridged gaps between alienation and belonging, creating common ground where judgment was left at the door. At first, sitting in rooms filled with strangers felt like stepping into a battlefield without a shield, but over time, it became a sanctuary where I learned the power of collective strength. The stories I heard reminded me of the complexity of addiction—not a personal failure but a systemic, societal affliction that ensnared countless lives. This understanding softened the edges of self-loathing and replaced isolation with connection. It was in these groups that I began to cultivate patience for the arduous journey ahead and gratitude for every small step forward.

Alongside these structured supports, mindfulness practices crept into my days, often quietly and unexpectedly. Moments of stillness, though rare and challenging, taught me to witness my thoughts without drowning in them, to observe cravings without surrendering. Meditation was not a neatly packaged escape but a raw confrontation with my own restlessness—a way of learning to sit with discomfort rather than run from it. Sometimes, it meant simply noticing the heartbeat pounding in my chest instead of panicking at its urgency. Gradually, these moments became anchors, brief respites where peace could manifest amid the chaos. They were the cultivation of inner space, allowing me to breathe through the storm rather than be swept away by it.

Establishing new relationships also played a critical role in my evolving routine. Reaching out, rebuilding old friendships with those who supported my sobriety, and cautiously forging new connections felt both terrifying and affirming. Loneliness had been one of the harshest punishments of my past life; breaking free from it required me to be vulnerable, to admit that I needed others, and to trust again despite the risk of rejection. The process was slow, filled with setbacks, but each positive interaction was a thread woven back into the fabric of my sense of belonging. These relationships, grounded in honesty and mutual respect, became a source of strength and accountability, reminding me daily that my life mattered and that I was not alone in this fight.

Work, too, slipped painstakingly into the mosaic of my routine. Holding a job was not just a source of income—it was a declaration of purpose, a way to reclaim dignity and self-worth fractured by years of addiction and violence. The struggle to maintain consistent employment was real and relentless: the exhaustion from sleepless nights, the weight of shame from past mistakes that employers sometimes reminded me of, and the constant anxiety that my internal battles might spill over into my professional life. Yet, each shift completed was a victory, each paycheck a symbol of progress. Work demanded discipline and allowed structure where chaos had once reigned unchecked. It provided a tangible reason to keep moving forward, to answer the call of responsibility, and to invest in a vision of myself that extended beyond survival.

Perhaps the most profound thread running through my new routines was the gradual rediscovery of self-compassion—a concept both foreign and necessary. For years, my internal dialogue had been a torrent of criticism, anger, and self-condemnation. Learning to speak kindly to myself, to acknowledge pain without blame, and to accept that healing was not linear was one of the hardest lessons I learned. This shift did not come overnight but emerged through countless small moments of grace—the times I caught myself before plunging into self-hate, the moments I forgave myself for a stumble, and the instances when I celebrated progress no matter how slight. Compassion was the foundation beneath all the routines, the soil that nurtured growth amidst the ashes of hardship.

Through these deliberate patterns—mornings marked by quiet reflection, journaling as catharsis, therapy as confrontation, support groups as sanctuary, exercise as reclamation, mindfulness as refuge, relationships as lifelines, work as purpose, and above all, self-compassion as balm—I began to reconstruct a life that could withstand the storms. The routines were not rigid chains but flexible frameworks within which I could learn, falter, and rise again. They were spaces in which I could face the relentless epidemics of trauma and addiction not as a defeated soldier but as a warrior cultivating resilience.

Some days were still heavy, the shadows long and consuming, but the patterns of my new life offered pathways through the darkness. Even when temptation slithered close, whispering the siren call of old addictions, the routines stood firm—anchors in a turbulent sea. I learned that sobriety was not a single victory but a tapestry woven of thousands of choices, moments of presence, and acts of courage. Each new routine was a thread in that tapestry, a stitch in the fabric of a life being reclaimed from the brink.

Establishing these patterns felt like a rebellious act against the chaos that had once ruled me. It was proving to myself that a future beyond pain was possible, that new definitions of strength could emerge through vulnerability and persistence. Day by day, breath by breath, I was building a refuge not only from addiction's grasp but also from the hopelessness that had once threatened to consume me whole. In the delicate balance between routine and spontaneity, between struggle and healing, I found a life slowly, painfully worth living—and in that discovery, the faintest light of redemption began to glow.

Inner Dialogue

The morning light slips through the cracked blinds, scattering jagged shards of sunlight onto the threadbare mattress where I lie tangled in sheets that smell of mildew and old sweat. It's the same shade of grey outside as it was yesterday, as it will be tomorrow, as if the whole city's caught in a loop, a monotony broken only by the distant noise of sirens wailing and the low rumble of traffic spilling into the dawn. The weight of the world hasn't eased—it has merely shifted position, pressing down heavier, settling on my chest like a stone I cannot throw off. I lie there, half-awake, half-drowning in memories, caught between two selves—one forged in the fire of survival, the other one, the fragile kid I forgot how to be, hidden somewhere beneath layers of scars and hardened resolve. The inner dialogue that begins in these early hours crawls through the cracks of my mind like a restless worm, probing soft places I thought were sealed shut forever.

I ask myself, Who am I now? Is this the man I was supposed to become, or just a shadow wandering in the wake of a broken past? There's a part of me that wants to scream, to shatter the silence with the rage accumulated from years of fighting to stay alive, fighting not to fall deeper into the abyss. But instead, there's only a hollow echo, like a whisper swallowed by empty streets. I turn onto my side and stare at the peeling wallpaper, tracing its cracked lines as if they are veins running beneath the skin of a city that never gave me a chance to be anything but a ghost. The habits I've formed over time are no better than chains wrapped tight around my wrists—cigarettes that scratch my throat with every drag, the reckless pacing of my mind as it leaps to the worst possible scenarios, the persistent gnaw of hunger for validation and peace I've long been denied.

Every breath feels like a negotiation, a silent truce with the demons that inhabit my bones. I remember the first time I chose escape over confrontation, the moment when I slid into that cold embrace of numbness, chasing that fleeting high that dulled the jagged edges of pain. Back then, it was a lifeline, the only relief from a world that talked in threats and bullets. But now, as dawn breaks and the city stirs, I am left with the residue—the hangover of broken promises to myself, the fractures in my identity that refuse to heal overnight. Yet, somewhere in the murk of self-reproach and regret, another voice rises, quieter but persistent. It whispers about survival not as surrender but as resilience, about the possibility that the man who wakes up here, in this tattered room, can still rewrite the narrative he's been handed.

I confront the reflection in the cracked mirror propped against the wall, a face carved by the years of hardship—lines deepened by sleepless nights, eyes shadowed with melancholy yet still flickering with flickers of defiance. The reflection looks back, and for a moment, holds no judgment, just a silent understanding. This is my battleground, the internal war zone where victories are measured not in grand conquests but in the small moments of peace that feel like revolutions in themselves. I think about the scars—both visible and hidden—and how they map the journey that brought me here. Each one carries a story of a choice made, a lesson learned, a survival instinct sharpened to a razor's edge.

Coping mechanisms have become my language of survival, even though they sometimes feel like betrayals. I lean on the ritualistic—the few habits that stitch my fracturing world back together, like the careful brewing of cheap coffee in the morning, the way I repeat certain prayers or affirmations that feel both hollow and healing at once. The street vernacular I once wielded like armor in gang circles now echoes in my inner narrative, reminding me of battles fought not just in physical streets but in the maze of self-worth and identity that I'm still navigating. When anxiety tightens its grip, I breathe through the rhythm of this internal dialogue, squeezing out space where hope can reside, no matter how faint or fleeting.

Days collapse into one another, and with each passing moment, the struggle to reclaim a sense of self feels like walking through thick fog—every step uncertain, every direction cloaked in doubt. Yet within this fog, flashes of clarity break through when I revisit the boy I once was—innocent, curious, yet already bruised by the world's brutality. It's a painful recognition to see that desperate kid mirrored in the hardened man, to acknowledge the parts lost or left behind. But it's also a part of my healing to hold both selves simultaneously, to allow vulnerability to seep into the cracks so resilience can slip through in return. The inner dialogue turns from judgment to compassion, the harsh words softened by the realization that survival was an act of courage and not a failure.

I trace back through the maze of daily struggles—the waking nightmares of cravings, the restless nights haunted by memories that refuse to stay buried, the endless cycle of guilt and shame that threatens to drown me. Recovery is a slow alchemy of patience and persistence, a painstaking reshaping of identity that demands confrontation with uncomfortable truths. Therapy sessions have become a battleground and a sanctuary, a place where I peel back the layers of defense to face the rawness beneath. The conversations with therapists and support groups crack open old wounds but also plant seeds of hope, urging me to rewrite the story from a new vantage point. Each session unravels another thread of denial and defense, exposing the skeletons behind my addiction, the social and personal fractures that fueled my descent.

Through this unraveling, I am forced to reckon with the systemic forces that collided with my personal failings—poverty that tightened like a noose, institutional neglect that turned empathy into indifference, and a society eager to condemn rather than comprehend. The inner dialogue becomes activism, a call to witness and voice the unseen struggles that go beyond my own. To speak for those shackled by the same cycles of violence and addiction, to unearth the humanity beneath the stereotypes and stigmas that have long silenced us. This new narrative is both empowering and daunting. It demands I carry both the weight of my past and the responsibility of my potential future, a balancing act that often feels precarious but ultimately necessary.

The love and vulnerability I once feared now emerge as strange allies in this ongoing battle. Memories of tender moments with the Canadian escort—fragile, imperfect, yet real—flicker like distant embers warming the cold corners of my heart. Our relationship, fraught with instability and raw emotion, taught me that connection is a battlefield too, where weakness is not defeat but a doorway to understanding. Those days of shared pain and fleeting joy revealed that beneath the hardened shell, the capacity for love remained stubbornly alive, challenging the narrative of isolation and despair that addiction had woven around me. In the labyrinth of my mind, these memories pulse like lifelines, complex but crucial threads in the tapestry of my recovery.

Homelessness, too, leaves its mark on the dialogue within. The nights spent on cold concrete, the cacophony of street sounds punctuated by distant cries, the bitter taste of loneliness and invisibility—all converge in an internal monologue saturated with both despair and grit. Survival tactics honed on the streets become as much mental maneuvers as physical ones, a constant negotiation between hyper-vigilance and the desperate yearning for normalcy. The street became a classroom, teaching lessons in humility and humanity alike. The struggle to maintain dignity amidst squalor comes alive in the quiet voice that keeps whispering, "Hold on. This is not the end." This voice, at times barely audible beneath the roar of memories, refuses to be silenced.

Each day, the choreography of coping continues—meditation on pain, measured breaths against anxiety, small affirmations whispered under ragged breath, the ritual of reaching out, even when hope feels thin as smoke. In moments of darkest despair, I remind myself that the fight for redemption is not linear; it is a spiral, circling back to old wounds with new awareness, each time carrying a different shade of healing. The dialogue within becomes a dialogue with the past, present, and future selves—a triad of acknowledgment, accountability, and aspiration. The battle is not just external but fiercely internal, unfolding in the quiet spaces where pain collides with possibility.

In this intricate dance of self-discovery, doubt frequently storms in, uninvited and relentless. What if I'm not strong enough? What if the demons claim me again? What if redemption is a mirage that keeps retreating as I draw near? These questions swirl in the recesses of my mind, threatening to undo the fragile progress made. Yet, the inner dialogue grows resilient by necessity, a bulwark built from the fragments of shattered dreams and fierce determination. It whispers not just of the pain endured but the lessons learned—how each scar, though painful, speaks of a survival instinct that refused to be extinguished. It insists that healing, though imperfect and painstaking, is attainable.

Hope, once a foreign concept, now resides in the punctuation marks of my thoughts. It glimmers in the gradual reclaiming of self—through therapy, advocacy, and the tentative rebuilding of trust, both in others and myself. The inner dialogue transforms from a litany of regrets to a narrative of possibility. I visualize the man I want to be, not as an escape from the past but as a continuation enriched by hard-earned wisdom. This new self is not unblemished but defiantly whole, a mosaic of broken pieces made resilient by the very cracks that tried to destroy him.

The day presses on, but within the shifting light of the room, a quiet revolution unfolds. The inner dialogue continues its ceaseless flow—a testament to the unyielding human spirit, wrestling with shadows yet carving out space for light. It carries me forward, one breath, one thought, one choice at a time, navigating the complex terrain between who I was, who I am, and who I strive to become. The journey is far from over, but in this moment, I stand on the threshold of a new chapter, the fragile yet fierce dialogue within echoing a truth I am only beginning to grasp: that redemption is not a destination but a relentless, courageous act of becoming.

Community Support

In the shadowed corridors of recovery, where each step forward often feels like wading through a dense fog of uncertainty and temptation, community support becomes not just a lifeline but a beacon—flickering sometimes, steady at others, but always present. For someone like me, whose history was carved out in the harsh, unforgiving streets where trust was currency thicker than money, learning to build connections with others in recovery demanded an entirely new form of courage. It wasn't about muscle or bravado anymore; it was about vulnerability—a strange, alien sensation to a guy who for so long survived by hardening every inch of his skin against the world. Every dawn was a battle, every interaction a test of resilience. But within this collective struggle, something remarkable began to take form: an unspoken fraternity born from shared pain, mutual understanding, and a common desire to reclaim a life that addiction had tried so hard to steal.

In the earliest days after my release from jail, when the weight of my own demons felt almost unbearable, I found myself drawn to community centers that offered recovery programs and support groups. These were places I'd never imagined stepping foot in while I was in the grip of active addiction. The sterile walls, the circle of chairs, the gentle but probing questions—all struck me as both alien and necessary. Initially, sitting in those rooms was like being a foreigner dropped into a country whose language I barely understood. People spoke about feelings—trust, shame, hope—words I'd long buried under layers of aggression and denial. But slowly, as I listened, something inside me softened. I realized that here, I was not alone. Others carried scars like mine, some deeper, some more fresh, but we were all wrestling with the same relentless tide of craving and regret. That gradual revelation, that recognition of shared humanity, formed the bedrock of the connections I began to build.

It's tempting to romanticize the idea of recovery communities as evenly lit spaces of unwavering support and encouragement, but the reality is far messier. These relationships were forged in an atmosphere heavy with skepticism and, at times, raw, unfiltered pain. Opening up about the nightmare of addiction—a nightmare that encompasses violence, loss, and self-destruction—meant exposing wounds that had long been kept hidden beneath a facade of toughness. Trust was elusive because it was so often broken. Early on, I saw firsthand how fragile this network of support could be. Relapses triggered doubt; moments of silence were sometimes mistaken for rejection. Yet, paradoxically, these imperfections served to deepen my understanding of what it meant to truly connect. I began to witness that recovery was not a linear path but a convoluted journey marked by stumbles, apologies, forgiveness, and renewed commitments. This gave me permission to be imperfect, to fail and try again without shame.

The daily struggles were immense. Even the seemingly mundane tasks—waking up sober, confronting cravings, navigating triggers buried in everyday life—required the strength that often felt beyond my grasp. But community support provided a crucible for learning coping mechanisms essential to survival. Group sessions weren't merely about sharing stories; they became workshops in self-awareness and emotional regulation. Techniques like grounding exercises, mindfulness breathing, and cognitive reframing were initially met with skepticism but soon became vital tools. More importantly, the reciprocal nature of these groups—where you both give and receive support—helped cultivate a sense of purpose that had long eluded me. Listening to others' pain and victories, and in turn bearing witness to my own, transformed self-pity into something more manageable, something I could face head-on.

One of the most profound realizations during this time was how the act of building connections in recovery became a mirror reflecting my own journey of self-discovery. The relationships I forged challenged my preconceived notions about strength and weakness. In the raw, unvarnished dialogues, I encountered my own fears and hopes refracted through the experiences of others. Seeing someone else grapple with their past or celebrate a month of sobriety sparked introspection that revealed parts of myself I'd never dared to confront. This communal reflection stripped away layers of denial, revealing the person I wanted to become beneath the scars. It pushed me toward embracing accountability—not just to others in the group but to the man staring back at me in the mirror. In those moments, community support was not merely external—it became intrinsically tied to my internal healing.

The complexity of these relationships cannot be overstated. On any given day, the mood within a recovery community could oscillate between hopefulness and despair. Celebrations of milestones were tinged with the sober recognition that addiction's shadow could reach around any corner. Watching a lifelong friend relapse was heartbreaking, a stark reminder of addiction's cruel grip. But equally, seeing those same individuals get back up, recommit to the struggle, and inspire others with their tenacity was a testament to the human spirit's stubborn refusal to quit. Within this mosaic of tragedy and triumph, the bonds of community support were knitted tight—sometimes with threads frayed but rarely broken.

That support extended beyond formal meetings into the messy realm of daily life. When isolation whispered lies in moments of weakness, phone calls from fellow survivors cut through the silence. When the streets called out with familiar sirens and shadows, texts and visits from new friends pulled me back into the light. This network became a web of accountability and encouragement, a lifeline anchored in mutual experience and understanding. I learned that recovery did not mean retreating into solitude but rather engaging with others who walked a parallel path. The social fabric woven in these communities was a stark contrast to the loneliness and hostility that characterized much of my earlier life. In embracing this new social reality, I was rebuilding not just my sobriety but my entire sense of belonging.

At the heart of this community lay a shared language—composed not only of words but of glances, gestures, and silences that expressed empathy and solidarity. These subtle forms of communication transcended the spoken word, often conveying more than any confession or story could. A nod of recognition, a tear wiped away, a hand lent in steadying balance all became acts of unconditional acceptance. In such moments, the veneer of loneliness cracked, revealing a profound interconnectedness. This intangible support strengthened me in ways therapy alone never could. It offered a lived experience, a collective heartbeat that reminded me continually: "You are not alone in this fight." The power of this realization cannot be underestimated, especially when the world outside often portrayed me as a lost cause.

Self-discovery within this context was neither gentle nor instantaneous. It demanded wrestling with the ugliest aspects of my past—my betrayals, violence, and failures—while mapping out a new identity that could exist beyond them. The community's role was to hold space for that transformation, to witness the unsteady unmasking of a man learning to reclaim dignity and hope. Often, this required confronting deep-seated shame and guilt: emotions that addiction had thrived on and perpetuated. Sharing these burdens lifted some of the weight, but also forced me to reckon with the roots of my pain. It became clear that empathy extended beyond others' stories; it was something I had to cultivate for myself. That process was as frightening as it was liberating, a daily negotiation of who I was and who I wished to be.

The emotional landscape within these communities was often raw and unpredictable. Tears were common—sometimes in despair, other times in relief. Laughter, too, emerged as a healing balm amidst the gravity of shared struggle. Bonds were reinforced through moments of honesty that shattered old masks and defenses. With time, I realized that these connections forged a new family—chosen not by blood but by shared experience and commitment to survival. This was especially significant given the fractured or absent familial ties many of us had endured. The support here wasn't about perfection but about persistence—celebrating every small victory, offering grace in moments of failure, and encouraging each other to keep moving forward even when the path seemed impassable.

One cannot overstate the importance of mentors and sponsors within these networks. These figures, often survivors further along in their recovery, provided guidance not through judgment but through understanding born of their own wounds. Their presence was a lighthouse, offering direction when the seas of contesting emotions and external pressures threatened to overwhelm. Their stories were living proof that redemption was possible, and their belief in me helped cultivate a fragile trust in my own potential. The mentor-mentee relationship was a crucible of accountability, a safe container which allowed me to test my limits and voice my doubts without fear of rejection. This dynamic encouraged growth in ways that solitary reflection or clinical therapy could not replicate.

The role of community support also extended into practical assistance, which proved vital in the early stages of recovery. Many individuals face not only the internal battle against addiction but the brutal realities of housing insecurity, unemployment, and fractured family support. Within recovery groups, resources were shared: places to stay, job leads, access to medical care, and legal aid—forms of help that rebuilt the external scaffolding necessary for sustained healing. There was a collective recognition that sobering up is only the first step; rebuilding a life requires tangible support and opportunities for empowerment. Being part of this system imbued me with a sense of responsibility—to give back, to offer the same support that had been extended to me so freely, and to become an agent of change within the larger community.

As the weeks stretched into months, the presence of these new relationships began to shift my internal narrative. The voice of self-hatred that once echoed relentlessly softened into a more compassionate inner dialogue. Community support was not a cure-all, but it was a powerful catalyst for change. It fostered resilience by reminding me that my worth was not defined by past mistakes but by the efforts I made each day to move forward. The shared stories of relapse and recovery provided a roadmap littered with both pitfalls and beacons of hope. In moments of despair, remembering those tales kept me tethered to the conviction that healing was possible. The community became a living testament that human beings are capable of transformation, even when the odds seem insurmountable.

Coping mechanisms developed alongside these relationships also evolved. Earlier, I had relied on substances to mute pain and silence the chaos within; now, I learned to channel difficult emotions into creative expression, physical activity, or spiritual practice. Group discussions revealed that these alternative outlets were critical tools for maintaining balance. Some found solace in writing or art; others in meditation or exercise; many in volunteering or advocacy. Personally, I began volunteering at local shelters and sharing my story with young people at risk. This gave new meaning to my past sufferings—transforming scars into sources of strength. The act of service reinforced my commitment to recovery and provided a belonging that extended beyond the immediate support group. It anchored me to a purpose greater than myself, one that invoked dignity and hope.

Yet, it was never a perfect or easy path. The fragility of recovery became evident in moments of sudden stress or unavoidable triggers: an argument with family, a bad day at work, or the mere sight of an old neighborhood. During these times, the support of my community pulled me back from the edge. The simple knowledge that someone, somewhere, understood what I was going through made all the difference. It dismantled isolation and replaced it with connection—the antithesis of addiction's lonely prison. This shift also required me to become more transparent about my vulnerabilities, another hurdle given years spent cloaked in emotional armor. But with each confession, I uncovered new layers of strength and acceptance, both from others and within myself.

Over time, the rhythm of life within this recovery community became a stabilizing force. Rituals evolved—weekly meetings, anniversary celebrations, check-in calls—that created a structure as vital as any medical treatment. This cadence nourished hope and built endurance, factors crucial for long-term sobriety. The community was not just a social circle but an ecosystem—each member's progress contributing to the whole. When one faltered, many rallied; when one rejoiced, many celebrated. This dynamic formed a reciprocal cycle of healing, reminding me that recovery was much less about solitary heroism and far more about collective resilience.

Looking back, the most profound gift of community support was the reclamation of my own humanity. Addiction had reduced me to survival mode, disconnected from empathy and love. Through these connections, I found a language of compassion that began within the group but soon spilled over into every aspect of life. I learned to listen more deeply—to others' stories and to my own inner voice. I discovered that healing demanded openness to pain and joy alike, a willingness to face both without flinching. These lessons transcended my addiction history and became foundational to who I was becoming. In the embrace of community, I found not just a path out of darkness but a route to reclamation and transformation.

The daily struggles, the coping mechanisms, the self-discovery that unfolded in the company of others—these were the elements that defined my journey through community support. It was a crucible of chaos and calm, despair and hope, brokenness and repair. And though the road remained long and sometimes treacherous, the connections I built became my compass and shield. In the fractured mosaic of recovery, these relationships were the spaces where I could be real, be raw, and ultimately, be reborn. Community support did not just help me survive; it helped me thrive. It taught me that even in the darkest corners of existence, when pain threatens to suffocate, the human spirit can find light—and that light often shines brightest when we walk together.

Chapter 10
Advocacy and Voice

Finding Purpose

In the beginning, I never imagined that my broken pieces could someday fit together to form something meaningful, something whole that could reach far beyond the streets where I'd first lost myself. For years, my days were swallowed by shadows—drug use, violence, and a desperate hunger for survival that numbed me to any thought of the future. But when those chains began to loosen, when the dust of my past settled just long enough for me to breathe again, a faint ember of something new ignited deep within. It wasn't just about saving myself anymore; it became about giving back, about using the jagged scars I carried to carve out a path for others still stumbling in the dark. The desire to create change, to be a voice for those silenced by addiction and violence, started small, almost imperceptible at first, but it grew fierce and unwavering, burning brighter with every story shared, every nod of understanding, every tear shed in communal spaces.

The genesis of this purpose wasn't born from some sudden epiphany or grand gesture, but from a series of moments so raw and real that they held no choice but to penetrate my hardened shell. My first public speaking event was far from polished. Standing in front of a cramped community center crowded with faces—some scarred by pain, others etched with exhaustion, many silently holding onto fading hope—I felt the old familiar dread gnawing at my insides. I wanted to retreat, to sink back into the safety of anonymity, but there was a flicker of something else pushing me forward. The words I spoke that night trembled and faltered, but they carried the weight of my past in every syllable. I told them about the streets that raised me, about the highs that made the world seem bearable, and the lows that nearly buried me alive. I spoke openly about addiction—the lies it tells, the friendships it steals, and the battles fought behind closed doors. The crowd was silent, weighing each confession, as if seeing a broken mirror reflecting their own struggles. Somewhere in that shared silence, I found the power of truth. It wasn't about me anymore, nor was it about shame; it was about connection, about casting light into the corners where few dared to look.

Community work followed naturally from that first step. I began volunteering at local shelters, counseling centers, and youth outreach programs, places where the scars of poverty, addiction, and violence writhed under the veneer of everyday life. These environments were often rough, fraught with skepticism and pain, but they reminded me of exactly where I came from. I met young kids with nowhere to go but the street corners I knew all too well, families fragmented by the same cycles I had survived, and individuals whose stories echoed my own in maddening new variations. Talking with these people, working alongside them, I realized that healing wasn't a solo journey—it was communal, messy, and profoundly human. It demanded vulnerability, empathy, and relentless honesty. It was in the small victories—helping a teenager resist the lure of gangs, comforting a mother mourning a lost child, advocating for harm reduction programs—that I saw the true meaning of giving back. With each interaction, I began to shed more of my own former shame and reignite the optimistic part of me that had been buried beneath years of pain.

As time went on, the platform I occupied expanded. Invitations came for panel discussions, town hall meetings, and conferences focused on drug abuse, criminal justice reform, and mental health. I learned to wield my voice not just for storytelling but for advocacy. In these spaces, I met allies and adversaries alike—policymakers hesitant to understand the complexities of addiction, community leaders torn between tough love and compassion, activists fighting tirelessly for reform. Navigating these arenas was challenging, often humbling; I was forced to confront the nuances of systemic issues that had once been faceless and abstract. The crack epidemic that had so profoundly marked my life was not merely a personal hell but a symptom of deeper societal neglect—broken policies, economic disenfranchisement, institutional racism—that required more than anecdote to address. It was here that I embraced a more deliberate role, becoming a bridge between the lived experience of those in the trenches and the pragmatic world of policy and reform.

The transition from survivor to advocate was accompanied by countless sleepless nights, reflections on painful memories, and the emotional labor of revisiting wounds that never fully heal. Speaking publicly meant more than recounting the past; it meant embodying hope and possibility, maintaining authenticity while offering a roadmap out of despair. At times, the weight was crushing—friends lost to overdose, xenophobic remarks from skeptics, and the constant reminder that, despite all efforts, the scourges of addiction and violence still thrived in many places. Yet, amidst the struggle, there were triumphs that reminded me why the fight mattered: a former gang member breaking the cycle thanks to mentorship, new funding for rehabilitation programs in neighborhoods once ignored, and testimonials from individuals inspired to seek help after hearing my story. These moments sustained me, transforming pain into purpose.

My relationship with the Canadian escort, though fraught and complicated, also became intertwined with this mission. Her own experience with vulnerability and survival deepened my understanding of human resilience and the necessity of compassion. Together, as we navigated the labyrinth of our pasts and the fragile present, she became an unexpected partner in advocacy, reinforcing the truth that healing is multifaceted and often requires the support of unlikely allies. Our conversations about identity, worth, and redemption enriched my narratives with a tenderness I hadn't known I could offer publicly. Through her, I saw more clearly that love and vulnerability weren't weaknesses but critical tools for transformation and connection.

As I continued to step into this growing role, the act of giving back morphed into something profoundly personal and political. Embracing advocacy meant challenging the stigma that shrouds addiction and criminal histories. It meant raising uncomfortable questions about how society reckons with brokenness and accountability. It meant urging communities and governments to invest in prevention, rehabilitation, and restorative justice rather than punishment alone. The stories I shared, the forums I spoke at, the grassroots networks I built—it all became part of a collective effort to rewrite narratives overshadowed by despair. The very streets that once seemed like inescapable traps were now the stage for renewal, testimony, and resistance. In this larger movement, I found a sense of belonging that transcended my former isolation, a purpose fueled by the belief that no one's life is beyond redemption if given the right resources and understanding.

Public speaking itself evolved from a fearful obligation to a sacred calling. Each time I approached a microphone, I took with me not only my scars but the hopes of many who had no voice or were unheard. The vulnerability required was immense, but the reward lay in the transformative power of shared stories. I learned to weave my gritty past with lyrical reflections—talking about the concrete jungle's harsh realities while also illuminating the flickers of light that persist even in the darkest alleys. Audience reactions ranged from stunned silence to tears, laughter, and applause, reminding me that connection is built on honesty and courage. These encounters taught me that the true essence of advocacy is not just changing policies or programs but shifting hearts and minds, dismantling prejudice one human story at a time.

In building community alliances, I connected with other survivors and advocates, forming networks that extended well beyond my neighborhood. Together, we crafted workshops, support groups, and educational programs aimed at youth and adults alike, focusing on early intervention, trauma-informed care, and peer support. The grassroots nature of this work underscored that change begins not only at the top but within neighborhoods, families, and individual lives. Each success, no matter how small—a young person choosing school over street life, a family reunited after years of separation, a policy initiative sparked by community pressure—reinforced the enduring value of persistence and collective effort. These achievements felt less like victories for me and more like milestones for all who refuse to let the past dictate their futures.

Even as I embraced this role, I remained grounded in the realities that birthed my journey. Addiction, violence, and trauma continued to ripple through communities, and my work often confronted these realities head-on. The juxtaposition of advocacy with ongoing personal struggle created tension but also authenticity. I could not speak about hope without acknowledging despair; I could not champion recovery without admitting relapse and vulnerability. This honesty smoothed the edges of perception—people saw me not as a distant figure who had "made it" but as a real person, flawed yet striving. This connection deepened the impact of my message and helped dismantle the stigma surrounding those grappling with similar demons.

Throughout this process, I discovered that finding purpose in giving back was not a destination but an ongoing journey—one that required continuous learning, humility, and adaptation. I immersed myself in education, reading about social justice, addiction science, and community organizing to better inform my work. I sought training in facilitation and trauma-informed care to serve more effectively. I embraced collaboration, recognizing that no one person holds all the answers. Each stage of growth brought fresh challenges but also renewed commitment. The road from darkness to light was neither straight nor smooth, but with every step forward, I reclaimed more of myself and contributed to rebuilding the fractured spaces around me.

Ultimately, the motivation to give back and create change is the thread that stitches together the fractured parts of my story. It transforms pain into power, loss into opportunity, and isolation into solidarity. Through public speaking, community engagement, and advocacy, I found a voice that once seemed lost—a voice capable of inspiring hope and mobilizing action. This newfound purpose became a beacon not only for me but for countless others navigating similar struggles. It is a testament to the human spirit's indomitable capacity to rise from ashes, to craft meaning from chaos, and to illuminate the future with experiences forged in the fires of hardship. In embracing this role, I didn't just find my redemption—I found a way to invite others into the light alongside me.

The Public Eye

Stepping into the public eye was like walking on a thin wire stretched tight between two skyscrapers, the city noises swallowing my feet, doing their shaky step dance. At first, the very idea of standing in front of others, sharing my darkest hours and raw scars, felt like a fractured mirror poised to shatter with one wrong move. But beneath the vertigo lurked a quiet hunger, the same hunger that had gnawed at me through those long nights in alleys and shelters, that whispered to me now to speak, to break the silence that had once cocooned me in shame and isolation. Advocacy was more than a role; it was a reclamation of voice, a defiant shout against the shadows that had tried to drown me. I wasn't just telling my story; I was holding a torch for others still stumbling through that pitch-black maze, a luminary guiding those hands searching desperately in the abysses of addiction and violence.

Each speaking engagement came wrapped in a tension that tightened my chest like the grip of a too-small glove. The room's lights often felt like interrogating suns, burning down on every pore, spotlighting my imperfections, every tremor in my voice, every flicker of doubt in my eyes. Yet, as I found my rhythm, stories began to pour forth with an authenticity that transcended rehearsed speeches. I spoke not as some distant survivor with a polished script but as a man who had walked through fire barefoot, who had felt the crushing weight of loss and the bitter sting of betrayal. The audience wasn't just listeners; they became extensions of my own community—faces of parents searching for answers to prevent their children's fall into addiction, young people who looked at me hoping for a tangible sign that recovery wasn't a myth, advocates and policymakers poised to reshape the broken systems. Their questions, their eyes gleaming with a mixture of hope and sorrow, became the forces that propelled me to dig even deeper, to unpack the nuances of my life lived at war with itself.

In the streets, where my battle had begun, advocacy took on an electric pulse, vibrant and raw. Working within the community demanded a different kind of courage—less about commanding a stage and more about kneeling in the dirt where the roots of pain took hold. I found myself meeting people where they were, not as some distant figure of triumph but as a living testament to the brutal cycle they were trapped in. Each conversation was a thread sewn into a tapestry of real lives, real struggles. Sitting on stoops with young men whose eyes flickered with the same desperation I once carried, talking through strategies that balanced reality and hope, I felt the full weight of responsibility. Sometimes those nights brought laughter, sometimes tears—but always the heavy iron of truth that change wouldn't come easy or fast. I witnessed firsthand the barriers of mistrust, of systemic neglect infected with decades of neglect, disinvestment, and cynicism. The cracks in their armor were wide, but so too was the potential for something greater to repair them if nurtured with patience and respect.

Navigating the labyrinth of community work also demanded a savvy that challenged my old instincts. The street vernacular and raw honesty I wielded didn't always translate in meetings with city officials or philanthropic organizations, who saw numbers on a chart rather than the bloodlines beneath. It required me to become a bilingual storyteller—shifting effortlessly from gritty, unvarnished truth to polished advocacy language, pressing the pulse points where policy intersected with lived experience. Walking into rooms filled with bureaucrats was a high-wire act where every word carried the weight of collective futures on the line. These interactions taught me the power of persistence. I learned to push beyond polite nods and hollow promises, crafting proposals that demanded resources for prevention, for rehabilitation, for healing. My own scars became an unwelcome but necessary commodity, evidence of the system's failures and a call to action to those who held the power to undo the damage.

But advocacy also unveiled the harsh glare of vulnerability—it stripped away all pretense of invincibility I once wore like armor. Every public appearance opened me up to scrutiny from far corners: the media eager to distill my narrative into sensational headlines, critics dismissive of redemption as a fluke. There were moments when I wrestled with the fear that my voice was too fragile to withstand the pounding drums of political indifference and social stigma. The lines between personal pain and public service blurred dangerously. I grappled with the emotional toll of recounting trauma repeatedly, sometimes feeling the old addictions stir in the shadows, lured by the lure of numbness. It was as if each retelling triggered ghosts, forcing me to relive knife fights, overdoses, and the suffocating silence of isolation. Yet, paradoxically, that repetition was also a form of therapy—a ritual of exposure that chipped away at shame and rebuilt my identity not as a victim but as a survivor and advocate. I found strength in the fragile threads of connection, in seeing faces shift from pity to understanding, in knowing that those moments might spark change in even just one life.

The rewards, though, were profound and often unpredictable. A single handshake after a talk, a note from a young person who found hope in my words, a community leader committing to tangible initiatives—these moments illuminated the darkness that had once seemed impenetrable. Advocacy became a living exchange, a reciprocal flow in which I gave voice and insight and received purpose and belonging in return. The transformative power of this work wasn't just personal redemption; it was the capacity to alter narratives imposed upon neighborhoods like mine, transforming statistics of despair into stories of resilience and possibility. That realization fueled my commitment even when setbacks threatened to crush my resolve. Long nights organizing outreach programs and advocacy campaigns, enduring bureaucratic red tape, or confronting skepticism in the halls of power underscored a simple truth: real change is never linear, never neat. It's messy, chaotic, born out of relentless effort and a refusal to let systemic neglect become a permanent legacy.

Throughout my journey as an advocate, I witnessed the kaleidoscope of human emotion and experience in all their complexity. I shared stages with former gang members who had turned their lives around, parents who lost children to the epidemic and now fought to save others, activists burning with righteous anger at broken promises. These encounters broadened my understanding of community beyond the narrow confines of my own story. I saw the shared threads that wove disparate lives together: the hunger for dignity, the yearning for safety, the fierce love that fueled survival against odds. In these spaces, physical and metaphorical, my voice became part of a larger chorus calling for justice and healing. The texture of advocacy work was textured—both exhausting and exhilarating, filled with setbacks that gnawed at hope but also glimmers that flickered like embers ready to ignite.

At the heart of my advocacy was the attempt to humanize the epidemic, to push past statistics and headlines to reveal the flesh-and-blood realities underneath. I wanted to dismantle stereotypes, those reductive images that painted us all as monsters or lost causes. By recounting the nuance of my existence—the moments of tenderness, confusion, and gradual awakening—I hoped to build bridges across divides, to foster empathy in spaces hardened by judgment. Advocacy was also an act of courage, forcing me to maintain honesty without self-flagellation, to balance vulnerability with resilience. In doing so, I stumbled upon an unexpected paradox: the very act of exposing my wounds became a source of power, a way to reclaim agency that addiction and violence had stolen. Each speech, each meeting, each community event was a brick laid in the reconstruction not just of my life but of the lives intertwined with mine.

Yet, the road was far from paved. Advocacy work exposed me to layers of exhaustion that physically and emotionally drained me in ways unfamiliar to my earlier struggles. The constant juggling of public demands and private healing was taxing; it forced me to develop thick skin without losing the sensitivity that made my message real. There were days when I questioned whether my words could cut through the noise of a society numbed by addiction's prevalence, hardened by systemic indifference. Still, the moments of breakthrough—when a policymaker nodded with genuine understanding, when a teenager's eyes lit up with a newfound will to fight, when a community found its voice—reminded me that advocacy was a long game, played out in small victories that when stitched together, formed a tapestry of hope.

Through advocacy, I also learned the vital importance of collaboration. Change didn't come from individual heroics but from forming bonds with those who shared the vision—social workers, outreach coordinators, health professionals, educators, faith leaders. Pooling knowledge and resources created a synergy impossible to achieve alone. I became part of networks dedicated to tackling the epidemic's roots from multiple angles: prevention, treatment, housing, and education. This web of allies anchored me and expanded the impact of my work. Together, we advocated for funding, shaped public discourse, and worked to dismantle the cycles that perpetuated addiction and violence. The camaraderie was a salve, a reminder that no one could walk this path solo and that deep connections spring from shared purpose.

One of the hardest lessons I confronted was in managing expectations—in myself and others. Healing and change were not linear; relapses, setbacks, and disappointments were part of the narrative. I had to learn compassion for those still trapped in cycles I had recently escaped and for myself when the weight of advocacy triggered old wounds. Balancing fierce hope with realism was an ongoing challenge; allowing space for failure without losing faith in progress was crucial. These tensions became themes in my talks—

honest reckonings that resonated because they mirrored the complicated realities many faced. Advocacy, I realized, meant holding multifaceted truths without seeking easy answers, embodying the arduous but rewarding path from brokenness to restoration.

Overall, becoming an advocate illuminated a transformation beyond the external milestones of sobriety and service. It was internal alchemy, turning pain into purpose, shattered identity into a mosaic of recovery and activism. The public eye no longer felt like a spotlight of exposure but a beacon of connection, a platform where my lived experience could ignite change in a world hungry for stories of resilience. My journey through advocacy was neither smooth nor simple, but it was profoundly alive—alive with the voices of many, including my own, rising in unison to remake a painful past into a powerful future.

Inspiring Others

Standing in front of a room full of faces—some weary, some skeptical, but all searching for a glimpse of hope—he felt a surge of purpose that cut through the haze of his own troubled past. The microphone was cold in his grip, a tangible anchor connecting him to the present moment, yet his mind wandered back to the nights when darkness seemed to swallow him whole, when the streets offered only danger and despair as constant companions. It was in that shared vulnerability that his story found its power, breaking through silence, shattering assumptions, and weaving threads of recognition and understanding among listeners who had known only isolation or judgment before. Each speech he gave was not merely about recounting the pain, it was about manifesting survival, about naming the monsters that once loomed so large and showing that they could be faced and overcome. The raw honesty of his words moved people into spaces they rarely dared enter—the messy intersections of addiction, violence, and redemption—and challenged them to reconsider the labels and stereotypes that had long defined him and others like him.

What started as hesitant talks at small community centers grew into invitations to schools, youth groups, and local forums where dialogue about gang violence and substance abuse was too often avoided or sanitized. He spoke not as a detached observer but as someone who had felt the jagged edges of survival strike deep into his soul. The echo of his voice mingled with stories of lost friends, moments of brutal reckoning, and flashes of hope that rose from the ashes of decay. He learned to read the room, sensing when a quiet nod or a misting eye marked a shift in consciousness. Some evenings were heavier than others, when a teenager asked pointed questions about choices or felt torn between the allure of fast money and the hunger for a better life. At times, his own reflection loomed large, trembling within the dialogue as he confronted the boy he once was and the man he was becoming. The ebb and flow of these encounters forged a rhythm for his healing journey—each story he shared added a brick to the foundation of a new identity, one rooted in purpose rather than pain.

The streets had been his crucible, but now community halls became his platforms for change. Beyond the testimonials and confessions, he embarked on tangible projects designed to reach deeper into the fabric of neighborhoods still grappling with cycles of violence and addiction. Partnering with local nonprofits, he organized workshops focused on conflict resolution and self-esteem, integrating life skills that had been absent from his own education. These sessions were never easy; often, participants arrived guarded or skeptical, their trust eroded by years of neglect and systemic failings. Yet, through patience and relentless commitment, small breakthroughs unfolded— glimmers of resilience ignited in young eyes, the first tentative steps toward self-reflection, or the breaking of silence that carried the weight of unspoken trauma. He treated every success, no matter how incremental, as a precious victory against the tides that had once pulled him under. It became clear that advocacy was not about grand gestures but about the persistent, everyday acts of connection and care that ripple outward in ways invisible yet profound.

The evolution of his role as an advocate ran parallel with his own continued battles with identity and forgiveness. He grappled with self-doubt and the fear of being defined solely by his past mistakes. The spotlight was a double-edged sword; while it offered a platform to influence, it also exposed the rawness of wounds not yet fully healed. Through this tension, he learned to embrace imperfection and vulnerability as strengths rather than weaknesses, weaving them into his narrative to dissolve the stigma around addiction and incarceration. This authenticity resonated deeply with audiences, many of whom found in his story a mirror for their own struggles or those of loved ones. His growing presence in media interviews and public panels enabled him to extend his reach beyond the immediate community, connecting with individuals and organizations across the country. The responsibility weighed heavily at times, but so did the conviction that every voice raised against apathy and ignorance contributed to the kind of social shift necessary for lasting change.

One of the most transformative moments in his advocacy emerged when he was invited to speak at a conference focused on criminal justice reform. Standing among policymakers, activists, and scholars, he offered a perspective forged in lived experience—a human face to statistics too often reduced to numbers. He recounted the systemic failures that funneled him from shattered neighborhoods to cemented cells, and the lack of support that deepened his despair rather than eased it. His words were unflinching but infused with hope, urging those with power to listen not just to the crime, but to the conditions and stories behind it. This opportunity expanded his vision of advocacy beyond personal recovery to structural change. He began collaborating with groups dedicated to rehabilitation programs, pushing for initiatives that prioritized mental health, education, and job training over punishment alone. It was a new battlefield, where the war was fought not with gunfire but with policies, empathy, and activism, and he embraced it with the same relentless grit he had brought to survival on the streets.

The feedback from those engagements often came in quiet, unexpected moments—a letter from a mother who found solace in his story, a former addict who credited his speeches with sparking the courage to seek help, or a young man who vowed to leave gang life behind after hearing the toll it took on this living proof of redemption. These connections reaffirmed that the journey from brokenness to purpose was neither linear nor complete, but rather an ongoing commitment to light in places long dark. His advocacy work increasingly incorporated elements of mentorship, offering guidance and hope to those at the precipice where he himself once stood. By sharing coping strategies, resources, and sometimes just unwavering presence, he built a network of mutual support that blended personal healing with communal empowerment. This organic growth of relationships underscored a fundamental truth: that recovery is not an isolated feat but a collective endeavor shaped by trust, compassion, and relentless belief in human potential.

Public speaking and community engagement also ignited a transformation in how he viewed his own worth and future. The stages that once felt alien and intimidating became spaces where he reclaimed agency over his narrative. He took pride not in avoiding his past but in confronting it head-on and using it as fuel for change. The process demanded ongoing reflection and emotional labor—revisiting painful memories, narrating failures as lessons, and modeling resilience without glossing over complexity. His voice matured into a bridge connecting disparate worlds—the streets he fled and the institutions seeking reform, the people enmeshed in cycles of violence and those striving to break them. By embracing this role, he no longer felt like a shadow of that old life but a beacon lighting the way forward. Each speech, each conversation, reinforced the profound ripple effect that one individual's courage can have in challenging the stigmas and systems that perpetuate suffering.

Throughout his advocacy, he maintained a steadfast belief that change required accessible and culturally relevant dialogue, especially for youth who often felt alienated by conventional messaging. He worked tirelessly to craft narratives that spoke their language—raw, honest, and laced with hope—while refusing to romanticize the harsh realities they confronted. His approach was grounded in empathy, never preachy, always nuanced, acknowledging the tangled web of choices, consequences, and circumstances. This authenticity created trust where skepticism had reigned and inspired young people to envision alternatives to violence and addiction. Witnessing their gradual openings—their questions, their tentative steps towards seeking help, their expressions of renewed self-worth—was the most potent affirmation of his mission. In these moments, the abstract ideals of advocacy crystallized into tangible human transformation, validating all the years of struggle that had forged his commitment.

The communal work expanded beyond speaking engagements to include direct involvement in organizing neighborhood events, outreach programs, and partnerships with local schools and law enforcement aimed at fostering dialogue and healing. He recognized that trust between communities and institutions was fragile and often fractured by generations of neglect and conflict. His role became that of a bridge-builder, facilitating conversations where fear and suspicion once dominated. It required patience and humility—acknowledging the failures and wounds of both sides while fostering a collective commitment to a safer, healthier future. This multidimensional engagement deepened his understanding of advocacy as a holistic endeavor—one that combined personal testimony with systemic critique and community collaboration. It was a stark contrast to the isolation and distrust that had defined so much of his early life, and it filled him with a cautious optimism rooted in shared humanity.

Through his growing influence, he also helped catalyze the establishment of support groups specifically tailored to former gang members and addicts, spaces where individuals could connect free from judgment and stigma. These groups became crucibles of healing, providing peer-led guidance and mutual accountability. He often facilitated sessions, drawing upon his own journey to model vulnerability and perseverance. His presence as someone who had walked through fire and emerged not unscathed but unbroken allowed others to envision paths forward. The power of these communal bonds reinforced the memoir's central truth—that no one heals alone, and that hope is nurtured in the company of others who understand. He continued advocating for expanded funding and recognition for such grassroots programs, emphasizing that recovery must be community-centered to be sustainable.

In embracing advocacy, he found his truest form of redemption—not as a distant victory but as an ongoing work shaped by every conversation, every handshake, every moment of daring honesty. His commitment to raising awareness about the intersections of gang violence, addiction, and social neglect became a rallying cry for empathy and action. He aimed not only to tell his story but to amplify the voices of those still caught in the shadows, weaving their experiences into the broader fabric of societal discourse. This public role was both a platform and a responsibility, one he carried with humility and fierce determination. As his influence expanded, so too did his resolve to remain grounded in the complexities of lived experience, rejecting simplistic narratives and instead advocating for nuanced understanding and comprehensive support.

Ultimately, the stories of impact he amassed were not trophies but testimonies—of young lives rekindled, of communities stirred toward healing, of policymakers prompted to listen and act. Each storytelling session was an act of solidarity, a beacon extended to those still navigating darkness, assuring them that their past did not have to dictate their future. The path was arduous and fraught with setbacks, but he carried forward with the profound knowledge that transformation was possible, embodied not just in his own survival but in the ripples of hope spreading outward through every life he touched. In this sacred work of inspiring others, he found the meaning and grace that had eluded him for so long—proving that redemption, once a distant dream, could become a shared reality.

Chapter 11
Family Reunions

Bridging the Gap

Reaching out across the years felt like navigating an uncharted labyrinth littered with the cracked shards of broken promises and bruised feelings. The gap between me and my family wasn't just measured in miles or missed phone calls; it was a chasm carved out by pain, betrayal, and silence that stretched wide and deep. When I finally summoned the courage to initiate the first tentative steps toward reconnection, I carried the weight of every word left unsaid, every hurt that had festered in the darkness, and the fear that perhaps some wounds were too raw to ever fully heal. The street vernacular that once defined my identity now felt like a barrier as much as a shield—was I still the same person my mother, my sisters, or my cousins had known? Or had addiction and years on the fringe changed me into someone unrecognizable? The process of bridging that gap wasn't linear, nor was it neat. It was jagged, messy, and absolutely essential.

Forgiveness became a double-edged sword. There were moments when I had to grapple with forgiving not only those who had hurt me, but also forgiving myself for the ways I'd failed my family. The memories of missed birthdays, phone calls ignored or cut short, and broken lives haunted me like shadows refusing to dissipate. I remembered how my absence had hurt my mother most of all—her voice trembling when she heard me on the phone after years of silence, the way her eyes lit up with equal parts hope and apprehension during our first face-to-face meeting after my release. I saw the lines of worry etched deeper into her face, the years of watching me spiral into chaos burning themselves into her soul. Yet, beneath that pain was an unyielding love, stubborn like the roots of an old tree breaking through cracked concrete.

Reconciliation wasn't merely about apologies exchanged or closures uttered; it was the slow relearning of one another's languages and expressions. It meant sitting in uncomfortable silences and confronting the rawness of misunderstandings that had ballooned into chasms of resentment. We had to learn to communicate again, often fumbling with words drenched in emotion, sometimes falling back into old arguments as familiarity bred tension instead of comfort. Each conversation was a delicate dance around fragile egos and guarded hearts, as years of hurt made us both defensive and cautious. I remember nights when, exhausted and vulnerable, I'd cry in my room afterward, haunted by fears that I'd irreparably damaged the bridges I so desperately wanted to rebuild.

Family dynamics had shifted too. Siblings who had once been confidants had become strangers, their own lives marked by scars and stories I was only dimly aware of. There was jealousy and mistrust to untangle, intertwined with worry over my still-precarious sobriety and the unpredictable nature of my past. While some welcomed me back with open arms and tentative smiles, others held back, their loyalty tangled with their own pain and skepticism. The eagerness to make amends collided with the need to prove that my change was real and lasting. It tore at me to see the guarded glances and cautious questions—"How do we know you won't fall back?"—because beneath those words was a fear I also carried.

Ongoing challenges relentlessly tested the fragile new normal we were trying to build. Old neighborhood ties and the temptation of old habits lurked at the edges of my progress, threatening to pull me back into the chaotic whirlwinds I had struggled so hard to escape. Every family gathering was a minefield of triggers and memories, a strenuous exercise in maintaining the hard-won sobriety while navigating an environment steeped in familiarity but also unresolved tension. Arguments sometimes flared, born from misunderstandings or the mismatch of expectations—sometimes from my impulsiveness when under stress or their lingering distrust. It became clear that rebuilding trust was not a matter of days or months, but perhaps a lifelong commitment.

Yet, in the difficult moments, there were glimmers of profound connection—unscripted laughter shared over old inside jokes reclaimed from the past like precious gems, the warmth of hugs that carried the unspoken words "I'm proud of you" and "I'm here." Slowly, the image I had carried of family as both sanctuary and battlefield began to shift into something more nuanced. I saw in my relatives their own vulnerabilities, their own scars and resilience. We were no longer just fragmented by division but intertwined through shared history and love, imperfect but unshakable. Healing wasn't magic; it was the steady drip of small moments that piled up—the morning coffee conversations, the careful listening, the setting down of weapons built from years of pain.

One of the most surprising aspects of this journey was how I had to relearn humility—not only to admit my mistakes but to accept help and express need without shame. I discovered that pride, once a defense mechanism that fueled my gang bravado, had become an obstacle to openness and growth. It required letting go of control, of the façade of invincibility, to show vulnerability to the very people I feared had forgotten me. Reconciliation demanded that I meet my family members where they were emotionally, not where I wanted them to be, which was as humbling as it was liberating. I was reminded that everyone carries their own burdens, and forgiveness is often as much a gift to oneself as it is to the other.

Navigating this reconnection wasn't just about mending personal relationships; it also spurred me to confront the larger systemic issues that had contributed to our family's fractures. Poverty, limited access to mental health care, and the insidious grip of addiction did not exist in a vacuum but shaped the lives and choices we had each endured. My attempts to bridge gaps within my own family illuminated the broader social chasms that demanded attention and compassion. This awareness ignited a new sense of purpose in me—to advocate not only for my healing but to be a voice for those who remained trapped in cycles of violence and addiction, whose families were still torn apart in ways I had painfully known.

The process of reintegration also forced me to confront my identity on multiple levels. I was no longer solely the product of the streets or the mistakes I'd made. I was becoming a man who could coexist with his past without being defined by it, who could offer love even after inflicting pain, and who could build a future even after losing so much. The concept of family expanded to include not only those related by blood but also those who had shown me kindness, forgiveness, and support during my darkest hours—friends, therapists, advocates—forming a new, chosen family that healed alongside the original one.

Of course, this path was not without relapses in both emotion and behavior, moments of frustration, and even regressions. There were days when the silence felt heavier than confrontation, when the wounds of the past threatened to reopen with a single misplaced word or glance. I learned to approach these setbacks not as failures but as inevitable parts of a complex healing process, requiring patience and resilience. Therapy helped me understand that reconciliation is a living, breathing thing—subject to change and growth, not a destination marked by a single event or conversation.

In the quiet moments of reflection, I often thought about what I wished my younger self could understand. That family, despite its flaws and fractures, carries a power that transcends pain—the capacity to nurture, to forgive, and to rebuild. I wanted him to know that even in the midst of chaos, there can be a return to the light, that bridges can be built even when the foundation seems crumbled. And most importantly, that redemption is never too late if one is willing to fight for it.

The journey toward reconciliation was undoubtedly one of the hardest I ever embarked on, fraught with uncertainty and heartbreak, yet it ultimately became one of the most transformative. It reshaped not only my relationships but also my perception of self, illuminating the way forward with a fragile but persistent hope. Bridging the gap was never about ignoring the past or pretending away the scars. Instead, it was about weaving together the torn threads of family, honoring both the pain and the love, and stepping into a future that, though imperfect, promised possibility.

Healing Wounds

The wounds we carry often run deeper than the eye can see, carving invisible scars into the heart and soul, twisting themselves into the core of our being. In a life marked by gang violence, addiction, and survival against impossible odds, healing those wounds is neither straightforward nor swift. It demands a brutal honesty, a relentless peeling back of layers wrapped in resentment, pride, and pain. Forgiveness, in this world, feels like a fragile and distant dream—sometimes even a betrayal to the suffering endured. Yet, it is only in the trembling possibility of forgiveness that any true peace begins to glimmer. The journey toward healing those deep-seated wounds is a tumultuous passage through the dark corridors of memory and emotion, where every step forward is fraught with the risk of opening old cracks that threaten to swallow you whole again.

For years, I carried resentment like a heavy stone in my chest, cold and grinding against my ribs, a weight that shaped my every breath. That bitterness was a fortress, shielding me from weakness but also isolating me from love, hope, and the possibility of change. It was easy to blame everyone else—the streets that swallowed my innocence, the gangs that promised belonging but delivered only violence, the family members who wavered in their support, the system that failed to protect, even the ones I loved who missed the subtle cries hidden beneath silence. But beneath that blame lay a more painful truth: a deep, aching sense of abandonment and betrayal that refused to be named. Resentment whispered that if I couldn't forgive, I could at least preserve clarity, hold onto the narrative of victimhood to justify the anger and the choices I made. Yet, this bitterness was also a poison slowly corroding what remained of my spirit.

The moment I first grasped that forgiveness might be a form of freedom felt like a paradox. How could I forgive those who had hurt me? How could I forgive myself for the havoc I inflicted on my own life? It was as if forgiveness demanded the impossible—a surrender of the rage that had powered my survival, a letting go of the very thing that had kept me moving. But in the silence of the nights, when the ghosts of the past felt loudest, I began to question everything. I realized that holding onto resentment was like clutching a burning coal, expecting it to roast the hands of those who wronged me, when in truth, it was searing my own flesh. The idea was simple in theory, yet profoundly complex in practice: forgiveness could not undo the past, but it could reshape my relationship to it, breaking cycles and opening space for mercy—to others, but most importantly, to myself.

Reconciliation was perhaps the most fragile step in this process, a dance between hope and fear. It required confronting the very people whose actions had left me scarred, facing raw wounds that had never fully closed. The process was not a neat closure or a sudden resolution but a series of messy, complicated conversations fraught with setbacks and silences heavier than words. Some relationships resisted repair, barricaded behind walls built by years of neglect, misunderstanding, and trauma; others, unexpectedly, bore tender shoots of renewed connection. Opening up to family members held a particular power—a recognition that we were all, to varying degrees, victims of a broken environment and imperfect choices. In some moments, tears flowed freely, pain shared rather than hidden, and I realized that the people I once despised were also struggling to grapple with their demons. It didn't erase the hurt, but it humanized it, slowly shattering the rigid lines I had drawn around love and hatred.

Yet forgiveness did not mean forgetting the past or pretending the pain never existed. Far from sentimental reconciliation, it was an acknowledgment of the complexity of human experience—the intersection of harm and healing, betrayal and compassion. Forgiveness was a deliberate act of courage, a daily decision to meet life with empathy and patience rather than with the defensive armor of blame and fear. Some days, the weight of my history felt unbearable, threatening to overwhelm every ounce of progress I had made. On others, a quiet peace settled over me, a whisper of grace reminding me that healing was not an event but a lifelong practice. Addiction had taught me the brutal lesson of how easily the mind could get trapped in cycles of despair, and now I saw that resentment was a different kind of prison: a mental cage forged with the shackles of unforgiveness.

In therapy, I often grappled with the paradox of vulnerability and strength, of opening myself to love while guarding fragile wounds. The therapist's office became a sanctuary where I could dismantle my defenses, piece by piece, in the company of someone trained to listen without judgment. Unpacking decades of anger, regret, and sorrow was exhausting and painful yet deeply necessary. I learned to sit with my feelings rather than bury them under numbing substances or explosive outbursts. Emotional honesty became a foundation for rebuilding trust—not only with others but with myself. Each session chipped away at the walls I had built, revealing a battered but resilient core capable of transformation. These moments of inner work were often overshadowed by the chaos of everyday life, but they were vital lifelines, reminding me that healing wounds required patience and unwavering commitment.

The challenges in healing run far beyond the individual psyche; they ripple through every aspect of existence, especially when the scars connect to systemic injustices and communal trauma. The environment I grew up in did not foster forgiveness; it demanded vigilance, aggression, and survival at any cost. To rise above that atmosphere meant confronting not only personal demons but also entrenched social conditions that perpetuate violence and addiction. I found myself torn between holding onto my past as a source of identity and breaking free from it to forge a new path. Advocacy became an extension of my healing—a way to transform personal pain into collective purpose. By sharing my story, I hoped to break down stigma, open dialogues, and create spaces where others struggling with similar wounds might feel less alone. This work illuminated a profound truth: healing wounds is not a solitary endeavor but a communal journey, one that thrives on connection, understanding, and shared resilience.

Even after years into recovery, fears would rise like tides, reminding me how easy it was to fall back into old patterns. The ghosts of addiction whispered cruel temptations; the memories of violence and loss lurked in the shadows. Forgiveness sometimes felt elusive, slipping through my fingers just when I thought I had grasped it firmly. There were moments when resentment flared unexpectedly— triggered by a glance, a word, or a memory—tugging at my guard like a wild animal. In those moments, I had to dig deep into the lessons my journey had taught me, to find strength in vulnerability and trust in the slow work of healing. It was never linear or neat. Each day presented a choice: to sink back into bitterness or to stand in the light of grace. The scars remained, but they no longer defined me; they became part of a larger story—one of struggle, survival, and ultimately, hope.

The process also demanded confronting the limits of forgiveness, the boundaries where healing must respect the reality of harm inflicted. Some wounds were so profound, the betrayal so raw, that reconciliation could never fully take place. Accepting this truth was itself a painful liberation. It meant letting go of the expectation that every broken relationship could be mended; it meant embracing imperfect peace and finding comfort in self-compassion. Forgiveness did not mean excusing, forgetting, or condoning destructive behavior, but rather choosing not to be imprisoned by it. In a sense, forgiveness became an act of self-preservation, a way of reclaiming agency in the face of powerlessness. This nuanced understanding was crucial to maintaining mental and emotional health while continuing to grow.

As the years passed, I began to notice subtle shifts in how I carried my past—the way memories transformed from jagged shards that tore at my spirit to worn stones, smoothed by time and reflection. The pain remained, but it no longer consumed my every thought. Forgiveness allowed me to hold the contradictions of my story—the darkness and the light—in a delicate balance. I learned that healing wounds was less about erasing pain and more about integrating it into a fuller sense of self, where compassion for others and for myself could coexist with the scars. This integration brought a new kind of strength—a quiet resilience that did not depend on denial but on acceptance and growth. It was a far cry from the hardened boy who once walked the streets with fists clenched and heart closed off.

In moments of solitude, I often reflect on how far this path has carried me. Healing is ongoing, a thread woven through every aspect of life—relationships, work, activism, self-perception. It is a fragile, evolving dance of progress and setbacks, triumphs and doubts. The wounds of the past shape me, not as chains but as markers of a battle survived. Forgiveness, once a foreign word, has become a guidepost, lighting the way through darkness toward a horizon of possibility. It has taught me that to heal is not to forget, but to transform, to reclaim the narrative from pain and use it to build a life rooted in courage, empathy, and hope. The journey is far from over, but each day brings a chance to choose healing over hurt, love over resentment, and light over shadow.

A New Foundation

Building a new foundation beneath the weight of old scars was perhaps the most daunting task that lay before me when I returned home. The cracks in our family's fabric, worn thin by years of neglect, addiction, and conflict, seemed as jagged and sharp as the streets that had swallowed my youth. Rebuilding wasn't about coating everything with fresh paint and pretending the stains had never seeped in; it was about peeling back the layers, raw and tender, and laying each brick of trust and understanding with painstaking care. I carried with me more than just memories of violence and despair—I brought back questions, regrets, and a desperate desire to mend the gaps where silence had long reigned. It was a collision course between the past and the present, a reckoning with wounds that refused to heal simply because time had passed.

The first conversations after my return were halting, brittle with unsaid truths and guarded by walls built from years of pain. My mother's eyes, once fierce and unyielding, now shimmered with a complicated mix of relief and guarded hope. The weight of our history hung between us like a thick fog—there were memories too sharp to hold without flinching, words once hurled in anger that still echoed in the chambers of our hearts. Forgiveness was not a one-sided act nor a quick fix; it was a marathon we ran together, step by trembling step. Sometimes it felt like we were circling a precipice, afraid to fall into the abyss of old resentments. Yet, beneath those layers of bitterness was an unspoken promise, a fragile seed of reconciliation seeking sunlight.

Understanding how to forgive, and just as crucially, how to be forgiven, required peeling away ego and defensiveness. I had to come to terms with the ways I had hurt those I love, knowingly and otherwise. Addiction had turned me into a thief—not just of goods, but of affection and peace. I'd stolen time, trust, and stability from my family, casting them into chaos. My mother's forgiveness wasn't instantaneous or unconditional; it was earned through my willingness to face those mistakes, admit them fully, and commit to change. I had to listen without interruption as she laid bare the nights she stayed up worried, the moments she wished she could have done more. In those moments, vulnerability was both a weapon and a balm. We shared tears, anger, and awkward silences that spoke louder than words ever could, each encounter a brick in the foundation we were trying to rebuild.

Reconciling was not a promise that everything would be perfect or simple. Old triggers lurked in conversations and unspoken expectations. Sometimes intense arguments would erupt, uneven and explosive, shadows of unresolved pain breaking through the fragile peace we fought to preserve. There were days seasoned with frustration when it felt easier to retreat into isolation rather than navigate the choppy waters of renewed connection. But each day was a refusal to let the past dictate the contours of our future. We began to learn a new language, one of patience and grace, where mistakes could be met with understanding instead of punishment. Slowly, a rhythm emerged—familial rituals like shared meals, unexpected laughter, and the quiet comfort of presence that began to knit us together in ways that mere words couldn't.

The challenge of building healthier family dynamics extended beyond my mother and me. Reconciling with my siblings, many of whom had grown up carrying their own versions of pain and resilience, was equally complex. Some held resentment for the fractured family we'd been, while others had distanced themselves to protect their own sanity. Engaging with them meant accepting their perspectives without the need for validation or defense. We began to talk in corners of the living room, avoiding crowded spaces where old patterns of competition and blame might resurface. I made it clear through consistent actions that I was committed not just to my own recovery but to the healing of our shared past. Trust, I learned, is not given lightly within families scarred by addiction and violence. It had to be earned repeatedly through honesty, reliability, and kindness.

In the process, the concept of forgiveness stretched beyond individual interactions and into the collective history of my family. Forgiving my parents for what they couldn't control or prevent— poverty, systemic neglect, and generational trauma—became a crucial step. My father's absence, once a festering wound, slowly began to make sense through the lens of his own battles and limitations. I came to understand that holding onto anger against him was a weight I was ready to put down, even if that didn't mean excusing his mistakes. The threads of empathy grew stronger as I recognized how pain had woven itself into all of our lives in different ways. With this understanding came a gentler approach towards myself as well—accepting my failures without allowing them to define or condemn me.

The new foundation wasn't just theoretical or emotional; it demanded tangible changes in how we lived and supported each other. Boundaries, once elusive concepts in our chaotic household, were introduced and respected, though not without teething problems. We established clear expectations, learned to communicate needs without judgment, and cultivated spaces where everyone could feel safe to express themselves honestly. This was not about erasing individuality but about creating a framework where every voice could be heard and honored. It was a painstaking process, akin to assembling a puzzle with pieces that didn't always seem to fit perfectly but gradually formed a meaningful picture.

The most profound shifts happened in the quiet moments, the everyday acts of solidarity and kindness that slowly chipped away at the hardened shells we'd grown into. Breakfast tables that were once scenes of tension became places of warmth, where sharing food also meant sharing stories, hopes, and laughter. Weekend visits transitioned from obligations into genuinely anticipated occasions. I found myself becoming an active listener for my family members, learning to witness their pain and triumphs without trying to fix or control, simply being present. That presence began to reciprocate in kind, weaving a delicate web of mutual care that sustained us even when challenges loomed large.

Of course, the journey was far from linear. Setbacks punctuated our progress, sometimes triggered by external pressures—a sudden illness, financial strain, or a reminder of past traumas. These moments tested our resilience and commitment to the new foundation we had built. When old wounds reopened, we grappled with feelings of helplessness and frustration, but each time, the choice to recommit was made consciously and collectively. We learned the art of apology not as a sign of weakness but as a testament to our humanity and shared desire for growth. Through these cycles of falling and rising, the bonds within our family grew deeper and more authentic.

One of the most challenging aspects was learning to forgive the smaller transgressions that, in a healed family, might be brushed off— forgotten promises, misunderstood intentions, moments of moodiness or withdrawal. Before, these would have sparked fiery tempers or icy silences that could last weeks. Now, they were met with curiosity and openness, inviting dialogue rather than harsh judgment. This shift required emotional maturity and patience, qualities that didn't come easily after years of chaos but grew steadily as we nurtured our new foundation. It was a testament to how far we had come that even in moments of disagreement, the underlying respect and love remained unshaken.

In parallel with the internal family work, we also began to navigate external relationships differently. Where once our family had been closed off, suspicious of outside involvement, we started to engage more with community resources and support systems. Family therapy sessions, though occasionally uncomfortable, offered a space for structured healing and understanding guided by professionals. These experiences helped us articulate feelings and patterns that had long been buried, challenging us to confront denial and defensiveness. The journey into family recovery was not solitary but often collaborative, drawing strength from shared vulnerability and the willingness to confront uncomfortable truths.

The presence of ongoing addiction within the family, for those relatives still struggling, created additional layers of complexity. Forgiveness in their cases included recognizing addiction as a disease while maintaining clear personal boundaries to protect our healing. This balancing act was delicate and fraught with tension—love and distrust coexisted side by side, creating emotional whiplash. We had to learn to tolerate uncertainty, to celebrate small victories without losing sight of long-term goals. These dynamics underscored how establishing a new foundation wasn't about achieving perfection but about embracing growth, imperfection, and resilience in tandem.

Cultural and generational differences further complicated our efforts. My parents' perspectives, shaped by their own upbringing and the societal forces they'd endured, sometimes clashed with my evolving worldview. Bridging this gap required empathy and a willingness to meet halfway, translating experiences and expectations with patience and care. Through heartfelt conversations, we explored the ways our histories intersected and diverged, using those insights to build bridges rather than walls. In doing so, I came to appreciate the rich tapestry of strengths and challenges embedded in our family lineage, understanding that healing was as much about honoring the past as it was about forging the future.

Even amid all the progress, feelings of guilt and shame often shadowed my steps. I grappled with the fear that I might never fully escape the recesses of my past, that my mistakes might forever mark the way my family saw me. Yet, each time I faltered, their acceptance—sometimes tentative, sometimes robust—pulled me back. Rebuilding our family was as much about breaking free from the chains of self-judgment as it was about forgiving others. It was a daily exercise in self-compassion, learning to speak kindly to myself and to accept that redemption was a process rather than a destination.

As trust deepened, the family began to share stories long hidden or forgotten, stitching together fragmented memories that offered understanding and context. My mother recounted moments of her own youthful struggles, painting a portrait less of a perfect matriarch and more of a flawed woman striving under impossible circumstances. My siblings offered glimpses of their own resilience, sometimes expressed in silence or in distant smiles rather than in words. These shared narratives became the threads that bound our lives anew, weaving a fabric richer and more enduring than any forced facade of normalcy could provide.

The reemergence of laughter and joy within the household was one of the most significant markers of healing. For so long, the sound of laughter had been a rare, almost forbidden melody, overshadowed by the constant hum of anxiety and conflict. Now, it began to fill corners and hallways, spontaneous and genuine. Humor, once a coping mechanism in solitary battles, became a communal balm, reminding us that even in brokenness, light could pierce through. It wasn't a denial of hardship but a celebration of endurance, a testament to the strength that stems from connection.

In the midst of constructing this new foundation, I also came to understand that family was not confined strictly to blood relations. Friends and mentors who had stood by me during my darkest hours became chosen family, integral to my healing and growth. Their unwavering support, honest feedback, and unconditional presence offered a parallel structure of love and accountability. Integrating these relationships into family life required openness and sometimes negotiation, but ultimately enriched the network of support that sustained my recovery. It reaffirmed that family is as much about who you allow into your life as it is about lineage.

The act of creating healthier family dynamics was also an act of reclaiming my own identity—no longer the lost boy wandering in the shadows of addiction and violence, but a man choosing to live with intention and integrity. Each moment of peace, understanding, or shared affection with my family was a brick in the edifice of my new self. It was in this redefinition that I found resilience not just to survive but to truly live. The foundation wasn't built once and left untouched; it needed constant tending and nurturing, like a garden that blooms best with patient, consistent care.

Ultimately, rebuilding the family was a profound lesson in hope. Despite the darkness that once engulfed us, the cracks did not define our story. Instead, they became the spaces where light entered—a metaphor for transformation born out of pain. We were no longer prisoners to the patterns of the past but architects of a future shaped by compassion, authenticity, and unyielding commitment. This new foundation, fragile yet steadfast, gave us all a chance to reclaim joy, connection, and purpose beyond the shadows that had long threatened to consume us. Through reconciliation and forgiveness, we forged a family that, while imperfect, embodied the resilience and love necessary to transcend even the deepest wounds.

Chapter 12
Crossing Borders

Leaving Home

The moment I stepped off the bus and into the cold, biting air of Canada, I felt the eerie hum of unfamiliarity press against my skin like a second layer I couldn't peel off. The city stretched out before me—glossy and laced with unfamiliar sounds, odors, and faces—and for the first time, I was far from the cracked pavement and crumbling stoops that had defined my entire existence. That world where I had carved my name into alley walls and memorized every dangerous corner was suddenly behind me, replaced by something cleaner, quieter, colder—but also strange and disorienting. The air was sharper, fresher, unlike the thick, humid smog that always clung to my hometown. The people moved differently, too; they didn't scurry or flash quick glances behind their heads. Their eyes seemed to hold a kind of calm I wasn't used to, but also a distant look, like seeing something and not quite recognizing it. In that foreign city, I was an alien, unsure of my footing on these slick, unfamiliar streets.

Canada wasn't just a country; it was a symbol of escape, yet it also became the mirror where I first confronted fragments of who I was beneath all the armor worn on those gritty streets back home. Walking through the clustered neighborhoods where the snow gently blanketed parked cars and silent schoolyards, I felt the tension inside me shift. It wasn't just distance from home—it was distance from the life I had known and the rage simmering within me, waiting to be unleashed or understood. The cold touched my cheeks harshly as I looked at the people in cafes huddled over steaming mugs, quietly exchanging words and kindness instead of threats. I wondered what it might be like to breathe freely in such warmth, without carrying the weight of constant vigilance. But there was also the subtle sting of being invisible and yet hyper-visible all at once—as a young Black man walking through a city that whispered foreignness in every language and accent. I was a wandering ghost in a land where I neither belonged nor fully understood the code of the streets.

Those early days in Canada played out like fragments of a dream—each moment filled with cool surprises and unexpected vulnerabilities. I found myself taking tentative steps into spaces I never thought I'd cross: art galleries where quiet, contemplative eyes met mine without judgment; parks where children and dogs dashed freely in the snow, laughter spilling like light; train stations pulsing with the lives of strangers who didn't nod suspiciously or ask questions with flashing eyes. I felt the first flicker of freedom here, but freedom was a double-edged sword. Without the protective walls of my neighborhood, without the chorus of gang names and street codes that had tethered me to survival, there was a scary spaciousness—a vastness in silence—that made me question everything I thought I was. It was in these moments, with the white breath puffing out in the cold night air and distant trains roaring beneath the city, that I began to wrestle with my own sense of identity. Where did I end, and where did the streets I had known begin? What was left of me without the chaotic pulse of the gang, without the fleeting answers crack gave me, without the constant fight to be recognized, feared, or respected?

Traveling beyond the borders I'd always known peeled back layers of my existence like an onion, making space for pain and memories that had been squeezed into tight, hardened compartments. The quiet hum of unfamiliar cities peeled back skin, exposing raw nerves. Canada became both a sanctuary and a testing ground. I walked beneath towering maple trees dusted in frost, their branches bare but resilient, mirroring the simultaneous emptiness and stubborn life I felt coursing through me. The chilly sun's pale glow cast a dim light on faces more gentle than the hard-lined visages back home. When I found myself alone in narrow coffee shops or wandering through snowy shopping districts, reflection flooded me. For a man used to the adrenaline highs and brutal lows of the crack epidemic's chaos, peace was disorienting and unnerving. Silence wasn't the absence of noise but a deafening presence I didn't know how to fill.

Everywhere I went, I was reminded of my roots, not just through pain but through the sharp contrast that placed my life's fractured story against what I saw as possibilities. I met people who had never glimpsed the taste of crack or trembling with the urge to pull a trigger. They wore their lives like a steady song, not an explosive beat. Yet, beneath their polite smiles and carefully constructed civility, I could see my own hunger—this desperate need to belong and to be understood, even if the language and customs were different. One evening, walking beside the frozen river under dim streetlamps, I caught myself wondering if the idea of "home" was tethered to a place, or if it was something more elusive. Could I find home in these alien streets, or was home a phantom I'd been chasing all my life, carried in my blood and bones rather than city blocks and familiar faces?

The visits to Canada forced me into uncomfortable reckonings with self. I had imagined leaving home would be an escape, that stepping outside my old life would mean stepping into light and safety. The truth was more complicated—more fractal, like catching shards of a shattered mirror and trying to piece together a reflection that made sense. I wrestled fiercely with questions of identity and belonging, attempting to reconcile the hardened youth from the crack-soaked alleys with the fragile man emerging in the quiet snow. I saw that belonging wasn't automatic but something to be claimed or sometimes created through deliberate action and vulnerability.

Canada exposed wounds and gave them room to breathe, even when I wasn't ready to confront every scar. It forced me to become both the observer and the observed, an outsider looking in and an insider wrestling with displacement.

There was a moment late one night, standing beneath a streetlight that flickered in the biting cold, when the weight of everything—the violence, the addiction, the loss, the hope—settled on me in a wave so profound it nearly bowed me over. I wasn't just leaving home; I was stepping into the unknown terrain of my own life's map, navigating not streets or turf but the unsteady geography of my heart and mind. The journey wasn't linear; it was dense with contradictions and shadows, glimmers and darkness tangled together. Each step away from home was also a step inward—a traverse through memories of laughter and pain, of broken promises and fierce love. Canada offered both distance and closeness, a paradox that echoed the paradoxes inside me.

In those foreign streets, wrestling with language and culture, I began to understand how trauma could hitch a ride across borders, how wounds refused to stay put and could shadow even the brightest of places. But I also saw the flicker of resilience, the stubborn pulse that had carried me through years of turmoil and could carry me yet further. The chilling winter nights carried a silent promise that even a soul as battered as mine could find warmth if it dared to seek it. I realized that sometimes leaving home meant not just running from pain, but moving toward a chance to rewrite the script—one draft at a time.

Traveling outside my familiar surroundings forced me to grapple not only with the practical challenges of movement but with the intangible pull of roots that won't be severed even by distance or desire. I carried my past within me like a shield and a shackle, while also learning that I could carry something new—a fragile hope, a thread of possibility woven into my broken narrative. Witnessing the quiet resilience around me in Canada taught me that survival wasn't just about brute strength or hustle; it was about patience, reflection, and a willingness to feel without pushing pain away. Home, I began to suspect, was less a place and more a state of being—something forged in moments of honesty and courage. And leaving home was the first, painful step toward discovering what that might mean.

The young man who wandered those snowy sidewalks was the same one who had learned to fight before he could walk, but here his fists unclenched slowly in the cold. He tasted the possibility that beneath the rubble of violence and addiction, there could be a tender, quiet life waiting for him. The bitter winds that night were both a shackle and a balm, urging him to face the parts of himself he'd long avoided. Canada became a crucible, not of escape but of transformation—where he first began to see that the shadows of his past need not dim the light of his future. And so, every mile traveled beyond the dusty streets of home was a thread woven into the fabric of becoming—a becoming not defined by where he was from, but by where he was willing to go, and what he was willing to fight for inside himself.

Encountering New Worlds

The city had always been a cage. Concrete walls, freight trains rumbling like restless ghosts, and streets woven in a patchwork of hardship. The air tasted of smoke and worn-out dreams, a heaviness that pressed down on every inhale. Leaving that world wasn't just about crossing physical borders — it was about stepping into an entirely alien reality. When I first set foot on Canadian soil, the shock was immediate, immense, and disorienting. It was like waking up in someone else's dream, where everything familiar was refracted through a lens I hadn't been granted before. The cold was sharper, crisp, and unforgiving, unlike the oppressive heat of the city's sweltering summers. But it wasn't only the weather that underscored the distance from home — it was the people, the rhythm of life, the unwritten social scripts that baffled me.

I remember arriving at the airport, the sterile corridors tasting of antiseptic with fluorescent lights that buzzed faintly overhead, so different from the hand-me-down bus stations and packed corners where I'd spent most of my youth. Beyond the gates, the world unfolded with a kind of quiet orderliness, people moving with purpose but without the charged tension of survival that always seemed just beneath the skin back home. Faces here were softer — kinder, maybe — but they also carried an unfamiliar reserve. In the city I knew, eyes met and held; in this new world, they flickered away, darting between polite acknowledgment and a subtle guardedness. It made me feel like an intruder in a privileged theater where the script was written without me in mind.

Navigating this strange society brought a profound cultural shock. Every small interaction became a site of negotiation and confusion — from ordering coffee at a café where the staff used formal greetings that felt awkward on my tongue, to grappling with conversations about weather, politics, or neighborhood woes, which seemed so distant yet oddly similar to the stories I carried in my marrow. There was a pervasive civility here that at first felt hollow, almost performative, especially when contrasted with the direct, raw, and often brutal exchanges of my old world. But as the days passed, I began to realize that this civility was a complex dance — a societal glue meant to hold together diverse communities under a thinner veneer of social distance. It wasn't perfect; nothing ever is — but it also exposed me to a new vocabulary of kindness and restraint I hadn't learned to speak.

One of the most unsettling yet illuminating experiences was the shift in how identity and belonging were constructed. Back home, identity was a code sewn into the fabric of where you came from, your crew, your turf, your survival strategies. It was a tribal vocabulary — a set of signs, slang, and rituals that broadcast allegiance and crafted a fragile shield against chaos. Here, identity was laid out differently. It appeared as a mosaic — pieces shaped by multicultural events, legal statuses, and an ongoing negotiation between heritage and the desire to integrate into something bigger, something more consensual and, paradoxically, isolating. I found myself caught in between. I wasn't fully accepted as "local," yet I wasn't completely tethered to my past either. I was adrift in a sea of hyphens and sometimes felt like a phantom slipping between the cracks of belonging.

There was an irony in that feeling — to become invisible in a land that prides itself on inclusion yet often falls short on genuine acceptance. I encountered moments where I felt less like a man rebuilding himself and more like a symbol of an uncomfortable difference. Some days, I wore the skin of my old self tightly, like armor against those invisibilities; other times, I tried to peel it back and absorb the new scripts, only to find the edges torn and bleeding beneath unfamiliar contexts. The challenge became learning not just the language of the streets back home or the formal English of the classroom, but the deeper, more subtle code of human nuance in this setting — the tacit rules about who gets invited in, who stays in the margins, and how people navigate complex intersections of race, class, and history.

Traveling across different parts of Canada only complicated this patchwork. The country is a vast stretch of different climates, cultures, and attitudes. Vancouver's rains blurred the city into a gray watercolor wash, where diversity thrived but sometimes felt like a show for tourists, an aesthetic more than a lived reality. Toronto hummed with urgency and a bustling ethnic tapestry, yet its shadows held neighborhoods where economic disparity registered as sharply as the old blocks I once walked. Montreal was a linguistic playground, its rhythms dictated by French and English, and a hint of European elegance clashing softly with North American grit. Each place peeled back more layers of who I was and who I could be — or might never be — in these new environments.

In one of my visits, I was walking down a street lined with the bold colors of a public mural depicting a mix of indigenous figures and immigrant faces, a celebration of resilience interwoven with a history marred by displacement and systemic struggle. I found myself stunned, mirroring the complex emotions that kept folding in on themselves inside me. The mural spoke to survival and renewal — themes I knew intimately — but it also raised questions about place and voice. Seeing symbols that intertwined stories both foreign and familiar nudged me toward a fragile connection with this land. It became clear that while my specific battles were unique, the human cravings beneath them — for recognition, dignity, belonging — transcended borders.

Learning to navigate was not just about external adjustments but also an internal reckoning. Memories spiraled in and out, sometimes fueling resilience, sometimes triggering despair. The loneliness pressed hard on nights when the apartments felt sterile and walls no longer whispered secrets of home. There were moments, too, when I watched other immigrants — men and women dressed in the half-confidence of outsiders — and recognized in their eyes a mirror of my own fractured identity. We were all wrestling with pieces of a past that refused to stay buried, with a future that remained hazy and uncharted. That shared struggle created small moments of kinship — a nod from a fellow traveler on this path, a momentary bridge across difference.

One afternoon, crisp with early fall's chill, I sat in a public park watching a group of children playing soccer — their laughter a melody that seemed to stitch time's fraying edges. I thought about how childhoods could so easily unravel when exposed to trauma, poverty, and addiction, and yet still kindle hope. It stirred something in me, a desire to not only survive but to understand and possibly influence these new worlds I found myself in. The belief that while I bore scars of an existence harsh and unforgiving, the lessons drawn, the empathy earned, had a purpose yet to be fulfilled.

Each encounter with those around me was a lesson in humility and curiosity. Conversations with neighbors who came from corners of the globe I only vaguely knew taught me humility about my own assumptions. There was Mehdi from the Middle East, who spoke of war with a quiet weariness but found joy in gardening; Aisha, a Black Somali mother navigating the chaos of raising six kids with a resilience that defied exhaustion; and Liam, a university student who struggled to reconcile privilege with an urge to make change. These encounters chipped away at my cynicism, softening my edges and expanding my capacity to believe in renewal not just for myself but for communities fractured by invisible wounds.

Language itself became a canvas for exploration. Despite English being my first tongue, the subtle accents, phrases, and slang here had their own cadence and flavor. I found myself adopting a softer tone, its inflections curving away from the harshness I once spoke, as if the very way I shaped words could affect my internal weather. Sometimes I stumbled, caught mid-phrase, wrestling with etiquette or humor that didn't quite translate, but those missteps were part of the learning curve — a gradual attunement to the rhythms of a place both close enough to hold but distant enough to unsettle.

This duality forged a new lens through which to view my history. The stark contrasts between my former surroundings and this new country did not erase the pain or complexity of my past; instead, they illuminated it from angles I had never considered. I came to understand that identity isn't a fixed monument but rather a river, sometimes muddied, sometimes clear, carving new pathways while carrying the sediment of old struggles. In this sense, my journey was less about escaping old ghosts and more about integrating them — allowing a fuller, more human story to emerge that could survive in multiple worlds.

Among the people who became part of this narrative was a Canadian escort whose presence was both grounding and bewildering. She embodied contradictions — strength and fragility, love and distance — offering a mirror to my own vulnerabilities. Through her, I learned that relationships here bore a different weight and texture, shaped by cultural nuances and the unspoken rules of emotional give and take. Our connection was a collision of past wounds and tentative hopes, a vivid reminder of how new worlds don't erase the personal inside stories but add layers to them, complicating and enriching the journey toward direction and purpose.

Each interaction, every small victory and misstep, stitched together a patchwork understanding that slowly transformed alienation into curiosity, and fear into guarded hope. I was learning to build new rituals — from coffee shop chats to quiet evenings spent reading in dim light — which felt like claiming a small piece of belonging. The process was slow, nonlinear, and often painful, but it was the only way forward. In this unfolding reality, culture was no longer a monolith but a living tapestry, constantly rewritten by those who dared to inhabit its margins.

In the stillness of long evenings, I often reflected on how these experiences reframed my understanding of home. It was no longer a singular place etched in memory or geography but a fluid concept infused with relationships, experiences, and the courage to stand in spaces where comfort isn't guaranteed. The crack epidemic, gang life, addiction — those chapters did not vanish but were remixed by a newfound capacity for empathy and learning. They became part of a larger story about human endurance and the possibility of change that crosses borders, both external and internal.

Ultimately, encountering these new worlds was a collision of pain and potential, pushing me toward an unexpected transformation. The cultural shocks, the estrangement, and the moments of fragile connection forged the backbone of a renewed identity — one not defined by survival alone, but by the active pursuit of healing and belonging. It was clear that the journey ahead would be long and complex, but the first steps — awkward, uncertain, and hopeful — had already been taken. Embracing this new version of self meant acknowledging not only where I came from but also the myriad paths that lay open before me, inviting exploration, growth, and ultimately, redemption.

Returning Changed

The rhythm of the road had always been a language I clumsily tried to speak but frequently misunderstood. But as I boarded the plane bound for Canada, a territory once as foreign as the distant dreams I'd long buried beneath the asphalt of my neighborhood, something inside me shifted. There was a quiet desperation in my departure, an unspoken hope folded tight into the creases of my worn-out jacket. Leaving behind the violent echoes of gunfire and the suffocating grip of addiction, my heart was heavy with the weight of survival, yet cautiously buoyed by an unfamiliar feeling— anticipation. The moments after takeoff were not lost on me. The city skyline, for all its glare and grime, grew smaller by the second, like a chapter I had read too many times, now folded away in the margins of my life. I leaned back, letting the hum of the engines carry me away from the streets I'd known and into the vast unknown, where identity was waiting to be questioned, reclaimed, or rewritten.

Every journey abroad had been stained with the complexity of my personal history, the shadow of my past trailing like a ghost that no passport could dispel. Canada came to me not only as a place but a symbol; a canvas far broader and less scarred than the one I had grown up painting on, stroke by jagged stroke. The bitter cold that seeped through the airplane window mirrored the chill of my internal solitude, but upon landing, the air was sharp with possibility. The cities—Vancouver, Toronto—each pulsed with rhythms unlike the ones etched into the concrete jungles I once called home. There I was, eclectically out of place but oddly at ease with the silence of snow blanketing the streets, muffling the chaos I had known. Wandering through the neighborhoods, I was confronted not with the cacophony of survival-driven violence but with patchworks of cultures coexisting in imperfect harmony. The stark contrast taught me that identity is not a fixed point but an amalgamation—both a tether and a prism— through which the world's harshness and beauty are filtered.

In Toronto, the multicultural bursts of street life reminded me that belonging is complicated and fluid. I had carried with me the burdensome narrative of being a product of the crack epidemic, a pawn in gang politics, a lost soul in addiction. Yet, here, amidst diversity and difference, I felt myself loosen the tight grip of self-condemnation. The city, with its sprawling neighborhoods each telling their own story, forced me to realize that my identity was not merely the sum of my mistakes or the scars inscribed on my skin. It was a living story, one that could stretch beyond the margins of my upbringing and the roles I'd been cast into. Walking down Queen Street, I caught glimpses of myself reflected in the faces of strangers— people forging forward through their own struggles and triumphs.

Those moments struck a chord, beating against the cage of my past, reminding me that redemption is less an arrival and more a perpetual journey.

The reality of traveling, stepping into these new environments, brought with it a painful contrast. There were times, walking past glass storefronts or sitting in cafes where the aroma of freshly brewed coffee filled the air, when the alienation was so sharp it cut deeper than any street corner knife. My accent, my skin, my history—all these traits that once helped me assimilate into the gang dynamics of Brooklyn now made me a stranger whose affiliations and allegiances were neither understood nor relevant. It was a humbling process, stripping away the bravado and survival tactics that had been my armor for so long. It was here, in these moments of isolation, that my identity fractured and reassembled, piece by fragile piece. I began to see that belonging was not something handed to me; it was something I had to forge, not from fear or violence, but from acceptance and openness—of myself and others alike.

Traveling also forced me to confront the nuanced layers of trauma buried deep within me. Standing in front of the Peace Tower in Ottawa, for instance, I was reminded of the relentless pursuit of order and discipline, ideals I had briefly glimpsed in my military days yet had difficulty fully embodying. There was a tension within me—a battle between the ingrained chaos of my past and the need for structure in the future I sought. The orderly streets and polite exchanges in Canada felt like a mirror to the internal order I craved but couldn't quite sustain. This dissonance fueled nights of restless reflection, where I wrestled with questions of who I was beyond the confines of my history, and what it truly meant to escape—not just physically, but spiritually and emotionally. The cold, the quiet, and the vastness of the landscapes stretched before me like a reminder that healing could be as expansive and patient as nature itself.

Each city I visited acted as a catalyst, a living classroom where I absorbed lessons about resilience, hope, and the contradictions of human nature. The kindness of strangers contrasted sharply with the distrust I'd honed from years of navigating gang life. The freedom found in the openness of foreign streets chipped away at the armor of suspicion I'd learned to wear. Yet, there was no simple balm, no instant transformation. Growth was insidious, creeping into the cracks of my defense mechanism, allowing light to filter in only slowly and with resistance. The journey home wasn't just a physical return but a metaphorical navigation back through these emotional thresholds. I was not the same man who had boarded the plane; I was neither fully healed nor entirely broken—somewhere in that liminal space lived a fragile hope that I could rebuild and redefine my existence.

My reflections on identity became a central theme amid these travels. The question "Who am I?" echoed louder with every passing mile, every new face, every unfamiliar street. Identity, I learned, was not a fixed label or a rigid allegiance. It was a fluid narrative constantly written and rewritten by choices, circumstances, and self-perception. Being away from home allowed me to see the patterns clearly—the cycles of violence, addiction, and alienation that had dictated so much of my early life. But it also revealed the undying ember of humanity beneath the ashes. The journey made me recognize that while my past would always be a part of me, it did not have to be my destiny. The act of traveling, immersing myself in new cultures and environments, became not just a means of escape, but a profound experience of self-reclamation. With each step taken on foreign soil, I reclaimed pieces of myself lost in the shadows of addiction and despair.

Moreover, the interactions I had in Canada shattered some of the stereotypes I held about the "other." The escort I had met earlier on my journey—the woman whose fragile bravery touched the corners of my guarded heart—had opened a window into vulnerability and love that transcended background or circumstance. Her own complexity defied simplistic narratives, just like mine. Together, our shared moments of tenderness and turmoil made me realize that every human is a mosaic of light and shadow, hope and fear. This understanding deepened during my time in Montreal, where the blend of languages, cultures, and histories echoed the layered, multifaceted nature of identity itself. The experience humbled me, teaching me that growth is not linear but tangled and complicated—like the very streets I had once roamed, seeking survival by any means necessary.

In a way, traveling to Canada served as a form of exile, a temporary removal from the world I had known, which was both necessary and painful. It was exile from the toxic rhythms of addiction and gang life, but also exile from the familiar, even when familiar had meant only pain and danger. I grappled with loneliness, the quiet absence of old friends who had once been inseparable parts of my daily existence, however destructive those bonds may have been. The silence of new cities pressed against me like a tight fist, but also offered a breathable space to confront demons long avoided. It was in this solitude that I discovered the strength to face the fractured truths of my history, to mourn losses that had been pushed underground, and to begin envisioning a future unshackled from the chains of automatic survival instincts.

Travel also underscored the universality of human despair and the powerful potential of resilience. Whether wandering the streets of Toronto or the snowy promenades of Quebec, I saw reflections of my own internal struggle in the eyes of others—people grappling with their own battles, masked with varying degrees of grace or desperation. It forged empathy where judgement had once lived, a critical step in my own healing. I realized the crack epidemic, gang violence, and addiction were not isolated phenomena but symptoms of wider systemic fractures. This awareness sparked a motivation to not only continue my personal recovery but to engage in advocacy and social change. The journey abroad expanded my world not just physically but morally and emotionally, urging me to use my story as a platform for voice and transformation.

Amidst these insights, the question of belonging remained complex and deeply personal. To whom did I belong, really? The answer refused simplicity. At times, it was the streets of my youth—the very place that crushed many beneath its weight—that called with an almost magnetic pull, a testament to the complicated loyalties I bore. At others, those foreign sidewalks and snowy walkways whispered of possibility and new beginnings. I understood then that belonging was less about geography and more about community, acceptance, and self-forgiveness. The journey showed me that one could belong to many places, yet find peace only by making peace internally. It's a paradox that held me hostage and yet offered freedom simultaneously.

Returning home was its own reckoning, a confrontation between past and present selves. I brought back with me the lingering scent of winter air, the imprints of distant streets, and an altered perspective that could not be unraveled. The neighborhood hadn't changed much; in many ways, the cycles of despair persisted with a bitter stubbornness. But I was different, the cracks within me now illuminated by the light of hard-earned insight. I carried the understanding that my scars, while permanent, did not define the totality of my existence. There was room for growth, room for hope. The journey had instilled a quiet courage, the kind that does not roar but whispers steadily, urging one forward even when the night seems endless.

In the end, travel became a chapter in my ongoing story of redemption—not an escape from identity but a profound excavation of it. The countless miles traversed brought me face to face with my fractured self, forcing me to reconcile the tumultuous history of my early years with the fragile promise of the present. It taught me that healing is an act of movement and reflection, a continuous cycle of letting go and holding on. The roads I walked in Canada were not just paths between points on a map, but symbolic journeys toward reclaiming dignity, understanding love, and embracing vulnerability. Through the lens of travel, I rediscovered the resilience that had carried me through darkness, the humanity I had once doubted, and ultimately, a more compassionate relationship with the man I was becoming.

Chapter 13
Reclaiming the Self

Education as Salvation

The streets had been my classroom for so long, their lessons brutal and unyielding, carved deep into the fabric of my existence with scars that refused to fade. Yet, somewhere beneath the hardened exterior that broke under the weight of loss and violence, there stirred a faint ember of curiosity, a yearning that pulsed quietly in the chaos—a hunger not for survival this time, but for something more sustaining: knowledge. When the world seemed lowered to raw instincts and fleeting moments of escape, the idea of education as salvation was more than just a cliché thrown around by textbooks or well-meaning mentors; it was a fragile lifeline, a compass pointing toward a horizon I hadn't dared to dream about before. I began to see that the streets had taught me how to fight and flee, but not how to soar beyond the narrow confines of the concrete jungle. For the first time, the pages of books called to me with a promise of liberation, and the classroom felt like a sanctuary—a place where rebirth was possible. The very texture of learning slowly seeped into my being, reshaping the rough angles of my world and peeling back the layers of self-doubt that had gathered like grime on my soul.

I remember the first time I walked into a community college classroom after my release. The air was thick with a different kind of tension, one laced with anticipation rather than fear. I was acutely aware of the stares; some suspicious, some indifferent, and perhaps a few admiring glimmers for someone like me stepping into a realm that once seemed utterly foreign. But rather than letting the weight of judgment tether me, I wrapped myself in the modest triumph of that moment—proof that I could inhabit spaces where I wasn't defined by the shadows of my past. Each lecture unfolded new territories: history, literature, and psychology. Subjects that peeled back layers of human complexity far beyond the narrow narrative I'd been trapped in. More than the information itself, it was the discipline of study, the rhythmic pursuit of understanding, that was transformative. I learned to sit still in my own mind, to wrestle with challenging concepts instead of running from pain through frantic motions. The pulse of addiction and survival quieted to a murmur as I started tuning in to the language spoken by scholars and dreamers.

But education was never a straight path—I wrestled constantly with the ghosts lining the corridors of my brain. Self-acceptance was a mountain I climbed one trembling, faltering step at a time. There were moments when the voices of self-loathing clawed their way back in, reminding me of who I once was, threatening to undo the fragile threads of progress I was weaving. It was easy to fall into the trap of thinking that my past mistakes made me unworthy of growth, that the mistakes of youth sealed the fate of the man I was becoming. Writing papers, reading theory, mastering new skills—it all rang hollow if the inner voice refused to forgive and nurture. But slowly, I learned to treat myself not as the sum of my worst days but as a mosaic of resilience stitched together by scars and triumphs alike. Education gave me the vocabulary not just to express ideas but to articulate the battle within. It offered a mirror to face my own reflection without flinching, to embrace not only my flaws but the fierce vitality that kept me pressing forward.

Career development became the practical embodiment of my intellectual awakening. As my confidence grew, so did my hunger to translate knowledge into something tangible, a craft or profession that would provide a foundation for stability and purpose. I enrolled in vocational courses, delving into fields that demanded hands-on skills and discipline, ways to earn respect and self-sufficiency beyond the precarious hustle of the streets. Coding, graphic design, mechanics— each module was a new world to dissect, learn, and master. With each completed assignment and earned certification, my portfolio became a testament to chapters unfolding beyond the painful narratives of my past. The work was grueling at times; old habits of procrastination and self-doubt surfaced like weeds, threatening to choke progress. But I showed up anyway. Each day's commitment was a brick in the foundation of a life I was building's gradually, fragile yet hopeful architecture.

Beyond the tangible skills, education planted seeds of empathy and awareness that spilled over into my personal growth. Through engaging with diverse perspectives and stories, I began unraveling the tight knots of animosity and bitterness that had hardened my heart. Learning about social systems, history, and psychology illuminated the systemic roots of the violence and addiction that had enveloped my youth. It was both a painful reckoning and an awakening. I wasn't just a victim or a product of poor choices but part of a broader tapestry woven with threads of inequality, trauma, and human frailty. This awareness fueled my resolve to become not just a survivor but a catalyst for change, equipped not only with personal agency but with the wisdom to advocate for others lost in the same darkness. Education became a bridge between the man I was and the man I aspired to be— one capable of not only surviving but leading, healing, and inspiring.

The relationship between self-acceptance and education was reciprocal and profound. As I fed my mind with knowledge, I nurtured my soul with understanding. I encountered stories that mirrored my own in unexpected ways, narratives of struggle and redemption that became beacons of hope. I discovered that self-acceptance wasn't a static destination but a continuous process of learning, unlearning, and reimagining my identity. My reflections evolved from harsh condemnation to compassionate curiosity—why did I make those choices? How did my environment shape my instincts? What wounds needed tending, and what strengths had I unknowingly carried all along? Education offered the tools to excavate these questions, to dissect them without fear, and to slowly reconstruct my self-image with grace and not guilt.

Social interactions in the educational environment further cemented this transformation. For the first time in years, I was surrounded by peers invested in their own growth, a community of learners who, despite their differences, shared the universal struggle of aspiration and doubt. Group projects and study sessions became arenas where I honed collaboration, communication, and patience— skills violently stripped away by years of isolation and conflict. Instructors challenged me not just academically but as a person, expecting integrity, punctuality, and resilience. Each meeting was a crucible that tested my commitment but also offered validation when I succeeded. Through mentorship and encouragement, I began envisioning possibilities that had seemed unattainable—a stable career, a voice in advocacy, a life defined by creation rather than destruction. The classroom became a microcosm of the larger world I was determined to re-enter on my own terms.

Yet, even as I advanced, there were setbacks—moments when the weight of my history collided with the demands of education and career-building. Stress could spiral into the familiar patterns of anxiety and self-doubt, whispering lies about my capacity to change. Old acquaintances tempted me with the siren call of past vices, exploiting moments of vulnerability. Financial strain was a constant shadow, threatening to derail my focus and stability. Housing instability sometimes siphoned my energy toward immediate survival rather than long-term planning. But I learned to lean into support systems I'd once scorned—therapists, support groups, community centers—resources that helped anchor me when the storm grew fierce. Education was no longer just an individual endeavor but a collective quest involving allies who believed in my potential. I began to understand that redemption was never a solo mission but a web of relationships, resources, and relentless hope knitted together by effort and grace.

Literacy, once a survival skill barely grasped, blossomed into a transformative power that I wielded with fierce determination. Reading became an escape hatch from the confines of my old world and an invitation into new realms of possibility. I voraciously consumed books ranging from biographies to social theory, poetry to autobiographies of fighters and thinkers who, like me, had navigated tumultuous journeys. Writing, initially a painstaking struggle, grew into a cathartic outlet—journal entries that traced my daily battles, essays that dissected social issues, creative pieces that imagined brighter futures. Each word was a thread in healing the fractured pieces of identity, a declaration that my story mattered beyond the street corners where it was forged. I discovered that sharing my voice was both an act of resistance and liberation—a way to reclaim narrative control and inspire others trapped in shadows.

The pursuit of knowledge reshaped not only the mind but the body and spirit. Routine replaced chaos: waking early to study, preparing meals with care, exercising discipline not only in mind but in health. These seemingly mundane rituals became anchors that reinforced a new rhythm of life—one driven not by survival but by growth and flourishing. I found myself setting goals tangible beyond the immediate moment: earning degrees, obtaining certifications, building a professional network, and nurturing relationships grounded in mutual respect and honesty. This path was neither straight nor smooth, marked by internal battles and external obstacles, but each milestone marked a victory over the narrative that once dictated I was bound to fail.

Perhaps most profoundly, education cultivated a newfound humility and patience that allowed me to view my past without hatred but with understanding. It helped me dismantle the destructive shame that had once threatened to consume me. In this process, I realized that knowledge itself was imbued with power—not only the authority to change external circumstances but the capacity to rewrite internal narratives. Redemption, I learned, was inseparable from the courage to face truth with compassion and the resolve to nurture the seeds of hope even in barren ground. This epiphany rendered the journey of education not merely an accumulation of facts but a profound transformation of being. I had moved from the shadows of the crack epidemic and gang violence into the light illuminated by learning, self-awareness, and resilience.

This light did not erase the darkness but illuminated a new way forward—a way that acknowledged my past without being imprisoned by it. It brought with it a sense of responsibility: the knowledge that my story could be a beacon for others lost in similar cycles. Armed with education, I found the voice to advocate for change, to reach out to communities battered by addiction and violence, and to challenge the systemic forces that perpetuated despair. This mission became inseparable from my purpose, an extension of the very salvation education had given me. It was a call to action etched deeply into my identity, where every lesson learned was a tool wielded in the fight for hope and justice.

Through education, I reclaimed ownership of my narrative—a journey marked not only by survival but by thriving, not merely by breaking cycles but creating new legacies. The act of learning became an act of defiance and healing, a celebration of the indomitable human spirit's capacity to transcend pain through knowledge and purpose. In the quiet hours of study and reflection, I found peace not by forgetting the past but by weaving it into a tapestry of redemption, purpose, and limitless possibility. Education was, indeed, my salvation—a transformative path that carried me from the shadows into a life newly illuminated by understanding, compassion, and the relentless pursuit of growth.

Career Steps

The morning sun poured through the cracked blinds of the small apartment, casting a pale, uncertain glow across the faded linoleum floor. It was a space that hummed with the echoes of past struggles—a silent witness to nights spent wrestling with demons both internal and external. Yet, this morning carried a different weight. It was one that hinted at possibility, at change not yet fully realized but palpably close. The journey from the chaos of streets and shelters to the slow, grueling, steady climb toward a meaningful career had been anything but linear. What had once seemed like a distant, unattainable dream now hovered within reach, though tethered to the hard truths and relentless persistence that lay ahead.

Finding meaningful work wasn't just about a paycheck or stepping out of unemployment; it had become a cornerstone of rebuilding a fractured identity. The scars left by addiction and gang life were embedded deep in the self, not just the body or reputation. Each stumble, each misstep in the labyrinth of drug-fueled nights and desperate street deals, had chipped away at any semblance of confidence. To forge a path forward meant more than learning new skills—it meant reconstructing a self-worth that had been eroded by years of abuse, loss, and chaos. The labor market was unforgiving for someone with a criminal record, a patchy work history, and the unmistakable aura of someone who had danced with death and come out bleeding. But the truth was, stability was not handed out as a reward; it was the product of grinding, relentless effort and an unapologetic belief in the possibility of redemption.

The first step toward this new chapter began humbly, often in places where the weight of past mistakes was felt most sharply. Community centers and job training programs became unexpected sanctuaries. Here, amid peeling paint and the scent of old coffee, there were mentors who saw past the stained paperwork and grit-streaked skin. Their challenge was not just to teach hard skills—computer literacy, resume building, customer service techniques—but to nurture the fragile shoots of confidence that fluttered in his chest. There was the bitter recognition that educational gaps had to be bridged; years lost to addiction had stunted growth, but not exterminated potential. Enrolling in GED classes was a critical milestone. The classroom was both a refuge and a battlefield; it demanded focus and discipline in a world that had taught distraction and instant gratification. Each lesson absorbed was a tiny revolt against the past, a declaration that learning could be a tool for survival and renewal.

Yet, education alone couldn't shield against the gnawing shadows of doubt. Self-acceptance surfaced as a more elusive requirement, its contours shifting with every internal dialogue. The stigma of addiction clung like a second skin, heavy and suffocating. Reintegrating into a workforce that often viewed him as a liability required peeling back layers of self-loathing to find resilience. The process of applying for jobs was riddled with moments of raw vulnerability; every rejection letter felt like a confirmation of unworthiness, yet each interview was a gauntlet to be endured with razor-edged patience. There were nights spent agonizing over what to say when asked about gaps or past offenses, carefully crafting a narrative that balanced honesty with optimism. Learning to frame the story was a skill in itself—sharing enough to be transparent without allowing the past to dictate the present.

Eventually, the turning point arrived in an unexpectedly steady role—entry-level work in a local nonprofit organization focused on community outreach and rehabilitation. The irony was not lost: a man once lost in the throes of addiction, now employed to guide others away from similar paths. The job did not come with a hefty salary or prestige, but it offered something far more valuable—a direct connection to purpose. The late nights in shelters, the raw stories shared behind closed doors, the faces weathered by hardship—they all fueled a renewed commitment to personal progress. Every interaction was an opportunity to reclaim dignity and rewrite the narrative, not just for others but for himself. Slowly, respect grew among coworkers and supervisors who observed in him a tenacity forged from fire and neglect. The work demanded empathy and endurance, but it also provided structure, an anchor in turbulent seas.

As confidence blossomed, so too did the appetite for further growth. The nonprofit environment, with its roots in social advocacy and community healing, ignited an intellectual curiosity that had lain dormant under years of survival mode. There were opportunities to attend workshops on case management, rehabilitation strategies, and grants writing. These experiences unlocked a dimension of possibility, revealing the vast spectrum of roles one could occupy in the service of change. The desire to contribute on a systemic level took seed. It became clear that career development was not simply a climb up the socioeconomic ladder, but a chance to build something enduring—a bridge between hard-earned survival skills and the power to influence policy and public perception.

With time, this growing engagement with advocacy sharpened focus on education once again, now with a goal toward formal credentials that might lend permanence to this fragile new identity. Returning to college part-time, juggling classes with work and the ever-present shadow of past traumas, required every ounce of resolve.

The curriculum was challenging, not least because it demanded confronting painful truths through sociology, psychology, and public health courses. Each class peeled back layers of systemic failure that had fed the epidemic that nearly consumed him. Understanding these dynamics provided both a theoretical framework and a renewed sense of mission. It was humbling to recognize how deeply entrenched the problems were, but also galvanizing to realize that personal transformation could evolve into collective action.

Still, balancing all these moving parts was a test of endurance. Financial pressures loomed large, forcing difficult choices between tuition payments, rent, and basic necessities. There were moments of near despair when the weight of responsibilities threatened to collapse the fragile equilibrium. Yet, this very tension underscored the imperative of persistence. The tiny victories—the passing of a difficult exam, positive feedback from a supervisor, an encouraging word from a professor—became lifelines that tethered him to hope. Slowly, the image of a 'career' shifted from a distant mirage to something palpable, shaped by patient effort and the courage to dream beyond survival.

One of the most profound lessons in this period was learning to accept imperfection—not just in the process, but within himself. The idea of a linear trajectory toward success was a myth; setbacks were inevitable and often painful. There were relapses, moments where the past's grip tightened unexpectedly, threatening to pull everything under again. Navigating these dark currents required a new toolkit, blending emotional honesty with practical strategies. Therapy and support groups remained critical, offering spaces where vulnerability was met with understanding, not judgment. These arenas nurtured growth, fostering a patience that was once alien. Self-acceptance was not static; it evolved daily as new challenges surfaced, demanding grace and persistence.

Through this continual work on the self, the narrative of worthiness seeped deeper into the bones, transforming internal dialogue from condemnation to cautious optimism. This transformation was mirrored in relationships forged along the way—mentors who saw beyond the surface, coworkers who valued integrity over history, and peers who shared the chaotic terrain of recovery and rebuilding. These human connections provided a rich soil in which confidence, responsibility, and ambition could take root. It was in these relationships that the protagonist found a reflection of the man he aspired to be, a living proof that growth was possible despite past damage.

Gradually, the career path evolved beyond survival-based roles into leadership opportunities. Invitations to speak at local events about his lived experience merged personal testimony with advocacy. Each public appearance was an act of courage and reclamation, stripping away layers of stigma and fear. The vulnerability inherent in sharing trauma with strangers was balanced by an overwhelming sense of purpose. The platform to influence others—especially youth teetering on similar precipices—gave voice to lessons harder won than any textbook could teach. This role as educator and advocate became integral to his identity, a powerful antidote to the helplessness once felt.

This commitment to giving back ignited a profound understanding of community resilience and the interconnectedness of personal and social transformation. Career development was no longer an isolated pursuit but entwined with a broader vision for social justice. The protagonist recognized that meaningful work was not merely about financial stability or climbing ranks but about impact— about creating ripples that could extend beyond individual success to spark collective healing. This realization energized his involvement in policy discussions, community forums, and collaborative initiatives aimed at dismantling the structures that perpetuated cycles of addiction and violence.

Equally significant was the quiet, internal work that shaped daily life—the dedication to self-care routines, the upholding of commitments previously inconceivable amid addiction's chaos. This stability forged through purposeful employment allowed for healthier habits, from regular meals to consistent sleep patterns, each a brick reinforcing the foundation beneath the rebuilding. These seemingly mundane acts became radical declarations of self-respect, underscoring the tangible benefits of having a steady paycheck and responsibilities that demanded presence and accountability.

At the same time, there remained a humility born of history. The path toward career development was always accompanied by the awareness of fragility—a precarious balance maintained through vigilance against complacency. This humility kept the protagonist grounded, fostering a work ethic rooted in gratitude and a fierce commitment to making each opportunity count. The scars of the past served as both caution and motivation, a constant reminder that progress required dedication and self-compassion in equal measure.

In striking this balance between ambition and acceptance, the protagonist cultivated a nuanced understanding of success—one measured not in promotions or salaries alone, but in the restoration of dignity and self-empowerment. The journey had transformed from a desperate bid for survival to an ongoing quest for meaning and contribution. This transformation radiated through every facet of life, from professional achievements to personal relationships, coloring the narrative with hues of resilience and hope. The career steps taken were more than mere milestones; they were emblematic of a life reclaimed from the brink, a testament to the indomitable human spirit's capacity to heal, grow, and ultimately thrive.

The long nights that once echoed with fear and desperation had gradually given way to mornings filled with purposeful routines and dreams reborn. Each job secured, each class completed, each skill mastered was a victory over the chaos that once seemed permanent. The journey remained unfinished, a lifelong work in progress, but with every forward step, the protagonist rewrote his story—not as a cautionary tale carved by addiction and violence, but as a powerful narrative of endurance, transformation, and the relentless pursuit of a life rebuilt from the ashes.

Embracing Identity

The journey to embracing my identity was less a straight path and more a winding road through shadows and light, a process that unfolded slowly, painfully, and ultimately with a quiet triumph that felt almost otherworldly. For the longest time, I wrestled with who I was—a tangled nexus of scars and dreams, flaws and hopes, mistakes and longings. It wasn't just about reconciling my past with the present; it was about weaving these fragments into something whole, something true. The man looking back at me in the cracked mirror was someone I didn't always recognize—someone marked by addiction's ghostly fingerprints, haunted by flashes of violence and loss, yet stubbornly clutching onto a flicker of ambition, a desperate need to find meaning beyond survival. Embracing identity was accepting not only the hard edges etched by pain but also the softer hues of resilience and growth that had quietly, persistently begun to color my life.

The first steps toward this acceptance did not come in any grand moment of clarity but through the slow accumulation of small victories in career development and education. After years lost to chaos, addiction, and incarceration, the idea of building a career once seemed like an impossible dream—an alien concept reserved for people who'd never been swallowed whole by street violence or addiction's dark abyss. But somewhere, deep under layers of self-doubt and shame, a voice whispered otherwise. I remember sitting in a community college orientation class, my hands trembling slightly as I filled out my registration papers. The classroom smelled faintly of old textbooks and dusty chalk, a stark contrast to the crack pipe smoke and gunpowder residue that had once permeated my lungs. The world I had stepped into was foreign, but there was something intoxicating about the possibility of learning, of acquiring knowledge that could shape a new narrative for myself.

Education became a salve, not just for my mind but for my bruised identity. It disrupted the internal script that had defined me as a "lost cause" or "just another statistic." Every book I opened, every essay I wrote, felt like reclaiming a piece of myself that addiction had obscured. The corridors of learning were, paradoxically, a sanctuary and a battlefield. I wrestled with my past in every classroom discussion and late-night study session. I had to confront the ugliness of my history with brutal honesty, not to glorify it but to understand its grip and unravel it thread by thread. There were days when the weight of self-judgment threatened to suffocate me—when the judgments of others, real or imagined, settled like stones on my chest. But persistence became my act of defiance against the internal demons that said I didn't belong.

Career development was the other pillar that fortified my evolving identity. My first jobs after leaving the military were humble, to say the least, but each one was a building block upon the shaky foundation I was constructing. Working in a community center, mentoring youth who reminded me painfully of my younger self, I found a purpose that transcended mere employment. It was not about climbing ladders of corporate success or chasing titles but about finding a role that allowed me to give back, to transform pain into empathy, and to use my experiences as tools for others' hope. The shift in perspective—from viewing my past as a burden to seeing it as an asset in advocacy and mentorship—was transformative. I began to understand that my scars were not just marks of failure but maps of survival.

Self-acceptance, though, was never solely dependent on external achievements. It demanded an internal reckoning that was often the hardest terrain to navigate. The quiet moments, alone with my thoughts, revealed the raw underbelly of identity—the fears of inadequacy, the lingering guilt over lives lost and mistakes made, the existential questions about worth and purpose. Therapy became a critical anchor in this process, a place where I could unspool the knots of trauma without fear of judgment. Through countless sessions, I learned to sit with discomfort and confusion, to listen to the parts of myself I had long silenced. It was humbling and terrifying, yet necessary. With each revelation came a deeper understanding: that identity is not a fixed point but a living, breathing continuum shaped by every choice, every scar, every moment of courage.

One of the most profound lessons in embracing my identity was learning to forgive—not just others but myself. The narratives I had clung to were often harsh and punitive, rooted in messages I absorbed growing up in a world where compassion seemed scarce. But forgiveness loosened those chains. It allowed me to breathe without the suffocating weight of self-loathing and to step into a version of myself that was flawed but worthy, broken yet healing. This shift was neither quick nor painless. Some days, old insecurities roared back with relentless intensity, and I had to remind myself that progress is neither linear nor unchallenged. Yet, over time, the cumulative effect of self-compassion began to reshape the lens through which I saw my story.

Embracing my identity also meant reclaiming the narrative around my cultural and familial roots, which were complicated and often fraught with conflict. Growing up in a community ravaged by systemic neglect, poverty, and violence, there was always an undercurrent of shame intertwined with pride—pride in survival and heritage, shame in the stereotypes and failures that external society would assign without understanding. I grappled with feelings of alienation from both my community and the broader society that often dismissed us as irredeemable. But through active engagement with cultural education, reconnecting with family members, and participating in community advocacy, I began to build bridges over that divide. Embracing identity was affirming my place not just as a survivor of hardship but as a bearer of a rich, complex legacy filled with resistance, creativity, and resilience.

This reclaimed identity revealed itself most vividly in the relationships I nurtured moving forward. The fragile connection with the Canadian escort, for example, was not just a chapter of vulnerability and love but a mirror reflecting the parts of me still learning to be seen and accepted. Through that relationship, I confronted the hesitancy to expose my true self—fear that opening up would invite rejection or hurt. But love, in its messy, complicated humanity, taught me that authenticity is the foundation of connection. I learned how to show up fully, with my scars and stories, and found that this vulnerability did not diminish me but made me profoundly human and relatable. The process of embracing identity was thus deeply intertwined with allowing others to witness and accept me, and in turn, learning to accept the imperfect humanity in them.

Professionally, this authenticity translated into a new mode of advocacy that was personal and powerful. Using my voice to highlight the realities of addiction and recovery, gang violence, and systemic failures became a way to legitimize my story and transform pain into purpose. Speaking engagements, community workshops, and written words became platforms where I no longer hid in the shadows but stepped into the light of truth. This advocacy work helped solidify the connection between identity and action—it was no longer enough to survive or to heal quietly. I had a responsibility to channel my experiences toward social change, to advocate for better resources, understanding, and compassion for those still struggling in the cycles I had once inhabited. Through this, the pieces of my identity—prodigal son, addict, soldier, survivor, advocate—began to harmonize into a coherent narrative of resilience.

The paradox of embracing identity was that it required both fierce honesty and profound gentleness. I had to stare down my darkest moments without flinching, acknowledge the damage wrought by years of addiction and violence, and simultaneously nurture the fragile shoots of hope and growth that appeared unexpectedly. The process was cyclical—moments of regression followed by leaps forward, periods of despair countered by bursts of joy and clarity. The narrative I was stitching was one of complex humanity, resisting simplistic labels or neat resolutions. It was in this messy unfolding that I found freedom—from the chains of my past, from the voices that sought to define me, and most importantly, from the internal critic that had dictated my worth for so long.

To embrace identity was also to confront the truths about systemic failures and social inequities that had shaped my life trajectory. I could not accept my identity fully without acknowledging the external forces—poverty, systemic racism, inadequate mental health resources, and relentless societal stigmatization—that contributed to my descent and complicated my recovery. This awareness deepened my empathy for others trapped in similar cycles and fueled my commitment to advocacy. It transformed my story from one of isolated struggle to a lens through which to understand broader social dynamics. Embracing my identity thus expanded beyond the personal to the communal and political, connecting individual redemption to collective responsibility.

In the quiet moments of reflection, I often marveled at the resilience it took to arrive here—the long stretches of loneliness on street corners, the nights vibrating with pain and fear, the times I felt as if the darkness would consume me whole. And yet, here I was, a man who had survived it all, piecing together a self from the remnants. The scars remained, yes, but they were no longer chains but maps: testaments to battles fought, to hell endured and overcome. The process of embracing identity was not about erasing the past but integrating it—holding the darkness and the light in one trembling, human hand.

Looking back, I realized that embracing identity was perhaps the most radical act of all—a reclaiming of agency in a world that too often strips it away from the most vulnerable. It was a declaration that I was not defined by my worst moments, nor by the judgments of others. I was a mosaic of every misstep and every moment of courage, every fracture and every attempt at repair. And in that acceptance, I discovered not only who I was but who I could still become. This truth carried me forward, into a future not free of challenges but illuminated by the light of possibility, redemption, and unwavering hope.

Chapter 14
The Rising Star

Looking Back

The past lies behind me now like a tattered map, weathered and torn by storms of pain and desperation, yet still carrying the faint outlines of a path that led me out of the deepest darkness. When I look back, it's impossible to separate the searing memories from the lessons they etched onto my soul. The shadow of those years—drenched in crack smoke, gunfire, betrayal, and despair—might have felt endless at the time, a black abyss swallowing every ounce of hope. But in retrospect, that harrowing journey was the crucible through which my spirit was forged. It's a strange sensation, reflecting on moments that once threatened to obliterate me, now standing as beacons of survival. Pain, I have learned, can either bury you or train you, and I chose, sometimes clumsily, sometimes stubbornly, the latter. The scars, both seen and unseen, remain permanent imprints—a testament not just to what was lost but to what endured and grew stronger beneath the surface. They remind me that the human heart, even when ground under the relentless weight of addiction and violence, has an uncanny ability to pulse with resilience.

I think of the boy I was then—fragile beneath a hardened exterior, craving acceptance in a world that too often saw me as disposable. To survive, I swallowed the bitter poisons of the street, seduced by the false promises of quick cash, fleeting camaraderie, and numbing escape. Every alley whispered survival stories, every shadow concealed deceit. The violence wasn't foreign; it was quotidian. Each gunshot, each scream etched rhythms into my waking nightmares. Yet, even amid chaos, there was an innocence — a flicker of unfulfilled potential — that refused to be snuffed out, no matter how often I tried to drown it in heroin, crack, and the adrenaline of gang allegiance. This duality of destruction and fragile hope is what defines those early chapters of my life. When I stare at those memories now, I don't just see a cautionary tale; I see a raw human struggle, an unvarnished saga of survival where every misstep was a lesson, every loss a painful punctuation mark in a long sentence that ultimately sought meaning.

As my addiction deepened, time fragmented into a collage of fleeting highs and devastating lows. Near-death experiences, raw nights trembling through withdrawals, moments of clarity swallowed by relapse — all fused into a chaotic symphony of self-destruction. The streets offered no solace, only bargains steeped in pain. Still, even amid the darkest nights, there were cracks in the façade—a glimmering shard of light waiting to be grasped. The turning point was not dramatic fireworks or sudden miracles but a slow, grinding awakening. That near-fatal shooting, that courtroom's cold judgment, and the sterile walls of incarceration stripped away illusions and forced me to confront the man I had become. Inside the silence of a cell, I could no longer run, no longer blame the world entirely. The journey toward redemption did not come in bursts but in painstaking steps—each therapy session, each support group meeting, a lighted candle in a cavernous darkness. Reflection resized my suffering from an unrelenting burden into a tool for transformation.

The military chapter added complexity to my story. The rigid routines and disciplined orders initially promised an escape from chaos but soon unmasked new internal wars. Military life taught me structure, responsibility, and the importance of camaraderie, but it also unearthed deeper wounds—trauma that festered beneath my steely exterior. I discovered that healing isn't linear; sometimes new battles arise even when old ones seem conquered. The fragile relationship with the Canadian escort opened another emotional door, showing me that love can be as complicated as addiction—capable of healing and hurting in the same breath. Vulnerability, once a ghost so terrifying I swept it under layers of bravado, now became a delicate bridge toward connection and empathy. These connections forced me to confront not just my past but my capacity for growth and intimacy, adding rich shades of complexity to my journey.

Homelessness was perhaps the bleakest season, when the street became both my prison and my school. Sleeping in alleys, scrounging for food, dodging dangers both external and internal, I learned the unforgiving lessons of invisibility. Society's blind eye turned me into a ghost, a silhouette on the fringes yearning for dignity. Yet even amid such despair, the ember of hope refused to die. Homelessness stripped me of superficial defenses but also offered clarity—the essence of who I was, beyond addiction and gang affiliation, laid bare and raw. It was in these moments of absolute vulnerability that the seeds of change were planted. Survival was no longer about drugs or violence but about reclaiming my humanity one breath, one step at a time. The slow crawl toward recovery began when I found myself willing to reach for therapy, to confess my darkest thoughts, and to accept help without shame. It was a jagged, uneven climb but an authentic reclaiming of self.

Recovery has been a mosaic of struggle and triumph, marked by gritty therapy sessions that dredged up painful memories and support groups where shared stories kindled solidarity. The journey demanded brutal honesty—facing not just the external demons but the internal ones, the shame, guilt, and deep-seated fears. There were moments when surrender felt like defeat, yet paradoxically, that very surrender was the bedrock of my healing. I've come to understand addiction not as a personal failing but as a battle against a formidable illness intertwined with systemic neglect and social trauma. This revelation fueled my commitment to advocacy, awakening a fierce desire to use my story not just for personal liberation but as a beacon for others trapped in cycles of violence and dependency. By speaking openly about my past, I sought to shatter stigma and build bridges where walls of misunderstanding previously stood. My work became an extension of my healing—transforming scars into tools for social change.

Looking back is not an exercise in self-indulgent nostalgia or remorse, but a deliberate act of reckoning. It is about tracing the intertwined threads of pain and perseverance that compose the fabric of my life. Each chapter of my past bears witness to the frailty and fierce resilience inherent in the human condition. Though the journey has been punctuated by moments of despair, it is also illuminated by acts of grace—offered sometimes by strangers, sometimes by family, and sometimes by my own stubborn will to survive. The darkness I emerged from was not a void but a crucible, shaping a new version of myself that chooses hope over destruction and purpose over aimless survival. I no longer view my history as a chain of failures but as a complex narrative where even my lowest points became pivot moments toward growth.

What lies ahead is not a promise of perfection but a commitment to continue walking the path of healing with honesty and courage. My gaze is firmly fixed forward, fueled by the knowledge that redemption is not a destination but an ongoing journey. Each day presents an opportunity to reaffirm my humanity, to make choices that align with my values, and to fuel the fire of resilience kindled by those who came before me and those who walk alongside now. I understand that the road to recovery is winding and at times treacherous, but it is also suffused with moments of unexpected beauty—in the laughter of new friendships, the quiet strength of self-acceptance, and the profound connection found in service to others. Purpose now guides me more than survival; the mission to uplift, educate, and advocate gives meaning that transcends my own story.

In this light, my past is both a cautionary tale and a source of profound strength. I carry the memories of loss and violence not as weights to drag me down but as brushes painting a nuanced portrait of redemption and resilience. The journey from the shadows of the crack epidemic to the light of hope is neither linear nor simple, but it is deeply human and relentlessly real. It is a testament to the truth that no matter how shattered a life might seem, the fragments can be gathered, pieced together, and transformed into a mosaic of meaning and possibility. Looking back, I am both humbled and emboldened—aware of the fragility that once threatened to consume me, yet buoyed by the unshakable strength that led me beyond it.

The essence of this reflection is not to glorify suffering or dwell in past mistakes but to reveal the unspoken resilience that underlies every struggle. My story is just one thread in a vast tapestry woven by countless others who have endured similar battles. With this perspective, the scars I bear become bridges to empathy, urging readers to see beyond stereotypes and fear, to recognize the complex humanity pulsing beneath headlines and hearsay. I hope that by sharing my voyage, I provide not only a mirror for those who find themselves lost in the darkness but also a lighthouse—a symbol that illumination is possible, even after the most harrowing nights. The promise of tomorrow, though uncertain, shines brighter when embraced with courage and an open heart to change.

As I walk forward, every step is an act of defiance against the forces that once sought to define me by pain and violence. The reclamation of my life is ongoing—a daily choice to honor the lessons of the past without being imprisoned by them. This balance, fragile and vital, fuels my advocacy and personal growth, reminding me that redemption is not reserved for a chosen few but is accessible to all who dare to confront their shadows and embrace their light. My journey stands as proof that from the depths of addiction, violence, and despair, it is possible to emerge whole again, not because the darkness was overcome without scars, but because the light was allowed, eventually, to shine through. Looking back, I see not an end but a foundation—a place from which the future can be built, strong and true, with purpose and hope lighting the way.

Sharing the Story

When I first started putting the pieces of my story into words, it felt like trying to hold water in my hands—slipping, elusive, impossible to contain all at once. But as the sentences began to form, as the memories calibrated themselves between shadow and light, I realized that sharing is where the true alchemy of healing begins. There's a sacred power in narrative, a transformative force that not only acknowledges the pain hidden in the darkest corners of experience but also rewrites the script, carving out new meaning from wreckage. My journey through the crack epidemic, through addiction's ruthless grip, the violence that tore through my world, and the fragile steps of redemption wasn't just a personal odyssey. It became something larger—a testimony to endurance, to complexity, to the indelible humanity beneath the layers of stigma and misunderstanding. The act of telling, of peeling back wounds with honesty, invites others into a vulnerable space that is at once fragile and fiercely resilient. It crafts a bridge between individual suffering and collective empathy, inviting connection where once there was isolation.

Narrative serves as both a mirror and a window. It reflects back to us the nuances of our own survival, the moments of desperation and courage that define who we are beneath the mask of circumstance. At the same time, it opens windows for others, allowing them to glimpse lives wrought with struggle yet flickering with hope. In the context of addiction and gang violence—the very systems engineered to silence, to fracture communities—storytelling resists the erasure of voices. It pushes against the numbing narratives of crime stats and news bites that reduce human beings to faceless numbers. By sharing my experiences, I wielded a weapon more potent than any gun or needle: my truth. And in that truth lies not just pain, but the raw material of advocacy and change.

Healing, I have learned, is never linear. It is a mosaic of steps forward and backward, of moments illuminated by despair and others lit by grace. Sharing the story gave shape to this mosaic. Each recollection, each confession, no matter how raw or ragged, built a scaffolding for recovery. Speaking about the crack epidemic's devastation, for example, meant confronting not just my own scars but the systemic failures that allowed such a crisis to escalate unchecked. It meant acknowledging the generational trauma inflicted on entire neighborhoods, the legacy of neglect and disenfranchisement that fuels cycles of violence and addiction. And yet, by naming these realities, I found a form of liberation. My story ceased to be a chain weighing me down and became instead a beacon signaling paths toward accountability and support. The vulnerability required to expose wounds is intimidating, but it invites reciprocal openness, fostering environments where healing is possible beyond the individual.

Advocacy through storytelling isn't merely about recounting hardship; it's a call to action—a vivid reminder that beneath social statistics and sensational headlines lie real people with real potential. When I speak of my past, I am not asking for pity but for understanding, for recognition that addiction and violence are symptoms of much broader failures in social infrastructure. This perspective shifts the conversation from moral failing to systemic crisis, demanding empathy over judgment. It urges policymakers and community leaders to listen to lived realities instead of relying on detached data alone. Advocates who carry stories like mine become crucial interpreters between worlds, those who translate the abstract into the intimately human, whose voices can sway attitudes, mobilize resources, and ignite reform.

Throughout my journey, sharing also forged unexpected solidarity. When I finally began to speak openly about the tangled web of my experiences—the gang ties, the near-death encounters, the military struggles, the fractured relationships—it was as if a veil lifted. Others who had traveled similar roads emerged from the shadows, their stories colliding with mine in a symphony of pain and perseverance. We became witnesses to one another's truths, breaking the loneliness that addiction often breeds. Storytelling transformed into a communal ritual of bearing witness, where survival became a collective endeavor. This sense of shared humanity reinvigorated my commitment to sobriety and service, reminding me constantly why redemption is not only an individual triumph but a communal responsibility.

Yet sharing the story is not without its risks. It demands a vulnerability that can leave wounds freshly raw, reopening chapters best left buried. The weight of exposing such intimate history—especially against the backdrop of stigma surrounding drug addiction and gang affiliation—can be daunting. Fear of judgment, of rejection, sometimes threatens to silence even the most determined voices. There's also the balancing act between authenticity and self-preservation: how much to reveal without being consumed by past shadows or retraumatized by public scrutiny. In navigating these perils, I discovered the necessity of trust—trust in supportive therapists, advocates, and community networks who could hold the truth with care and without condemnation. This infrastructure of support undergirded my ability to keep telling, to transform what was once a source of shame into a source of strength.

Moments of public speaking, writing, and advocacy became stages where I reclaimed power from pain. Each time I recounted the night I nearly lost my life to gang violence, or the suffocating spiral of addiction that seemed inescapable, it was as if I wrangled a bit more control over what once felt like chaos. The raw sincerity of these accounts could arrest audiences, dismantling preconceived notions and stirring compassion. But beyond stirring empathy, they served as warnings, as maps marked with the pitfalls to avoid, and the possibilities for renewal. There is unparalleled potency in hearing a lived narrative emerge from the margins; it humanizes policy debates and grounds abstract reforms in tangible realities. I could see glimmers of this impact when attendees afterward approached me, sharing their own struggles, offering gratitude, or expressing newfound understanding. Such moments reinforced the ripple effect of storytelling, where one voice can catalyze a chorus of healing and action.

Looking ahead, the responsibility that comes with sharing my story has become both a mantle and a mission. The trajectory from the crack epidemic's engulfing darkness to the promise of redemption is not unique to me, but part of a broader tapestry woven through countless lives. Narratives like mine can fuel critical dialogues on race, poverty, addiction, and criminal justice reform. They become instruments to dismantle harmful stereotypes embedded in media and cultural discourse—the myth of the irredeemable addict, the demonized gang member—replacing them with nuanced portraits of brokenness and resilience. In doing so, these stories foster environments where policies evolve beyond punishment and neglect toward compassion, rehabilitation, and investment in community healing.

There's an unspoken covenant that accompanies the act of sharing: to honor the fullness of the journey, with all its messiness and contradictions. It means standing not only in the triumphs of recovery but also in the humility of ongoing challenges. It demands constant self-reflection to ensure the story does not harden into static identity but remains a fluid testament to growth and possibility. This dynamic tension propels me forward, motivating a continuous re-examination of purpose and priorities. Sharing has thus become a lifelong dialogue—between past and present, pain and hope, individual and collective—that fuels both personal stability and broader social transformation.

The power of narrative also thrives in its communal dimension—through forums, support groups, workshops, and public engagements where stories intersect and multiply. These shared spaces become incubators of solidarity and innovation, where survivors and allies strategize together, building networks that transcend the isolation of addiction projects. Within these forums, the story is no longer a solitary thread but part of a vibrant fabric, stitched from diverse experiences and perspectives. The fabric yields strength unimaginable for any single individual. Witnessing this dynamic, I grasped that my purpose extends beyond telling alone; it includes listening deeply, amplifying other voices, and nurturing a culture where the exchange of stories is a central tool for healing and advocacy.

In embracing this collective narrative, I came to understand the interdependence of healing and activism. Recovery reinforced my voice; advocacy gave my recovery meaning. The stories of pain I shared prompted me to confront systemic violence and neglect with renewed urgency, transforming personal redemption into a platform for social justice. Each testimony became a building block toward dismantling the cycles of addiction and violence that ensnare so many. Simultaneously, as advocacy initiatives gained momentum, they fed back into my ongoing healing, providing purpose and a sense of belonging that had been elusive for years. This symbiotic relationship between storytelling, healing, and advocacy is, I believe, the cornerstone of meaningful transformation.

Sharing my story has also reshaped the very way I view identity and self-worth. No longer do I see myself solely through the prism of past mistakes or trauma, but as a living testament to possibility. The narrative acts as a container for complexity—acknowledging darkness without being consumed by it, celebrating strength without denying vulnerability. It allows for the coexistence of pain and hope, confession and courage, brokenness and rebuilding. In this space, identity evolves from victimhood into empowerment, and survival into thriving. It is a profound revelation that has freed me from the suffocating confines of shame and opened up boundless horizons of purpose.

Ultimately, the importance of sharing the story lies in its capacity to transform isolation into connection—a vital antidote to the loneliness that addiction, violence, and trauma impose. When people tell and receive stories, they weave threads of humanity that intersect across differences of background and circumstance. This weaving creates a collective narrative that is richer, deeper, and more compassionate. It is in this communal fabric that genuine healing takes root and advocacy gains momentum. The stories transcend the individual, becoming a beacon for those still struggling in the shadows, illuminating paths toward recovery, dignity, and hope.

So as I look forward, I carry with me the knowledge that my story is more than memory—it is a living force for change, a catalyst for understanding, a testament to resilience. I am compelled to keep telling it, not for self-aggrandizement, but for the countless others whose voices remain unheard. Sharing becomes an act of defiance against stigma and silence, a refusal to let past mistakes define the future. It is the light pushing back against the darkness, a declaration that no matter how grim the night, the dawn holds the promise of redemption. Through narrative, we find not only our own salvation but the power to help heal our communities, to inspire transformation on both intimate and systemic levels. This is the gift, the responsibility, and the profound hope embedded in the simple but profound act of sharing the story.

A Brighter Tomorrow

There is a certain kind of quiet that settles over a soul when it's been dragged through enough shadows; a deep, reflective stillness that carries both the echoes of past storms and the whispered promise of dawn. In those moments, I often find myself drifting into the spaces between memory and hope, wrestling with the person I was and the person I strive to become. It's a tension that's both tender and raw — like a wound that's beginning to heal but still throbs with the pain of what was lost. Standing here, at this crossroads in my story, I recognize now more clearly than ever that the path forward—my path—is one paved not with denial or escapism, but with a fierce commitment to growth, understanding, and compassion. I see that every step I've taken, no matter how dark, was necessary to bring me to this place where light feels not just like a distant glimmer, but a tangible reality I can reach for and shape.

I think back to those early days when survival was all I knew — when the world appeared as a battleground where trust was scarce, violence was constant, and hope was nearly impossible to grasp. There was something brutal and defining about that existence. The crack epidemic didn't just ravage my community; it tore through the very fabric of my identity, fragmenting my faith in humanity and in myself. I wandered through years marked by violence, addiction, desperate choices, and broken relationships. Each day was a fight against the dark that threatened to swallow me whole. But if I listen closely, beneath the chaos and despair, I hear the faint heartbeat of resilience — the unyielding pulse of a spirit refusing to be extinguished.

Now, weighing the balance of those fragments, I realize that my story is not just about falling apart but about the painstaking process of gathering those pieces and fitting them together into something whole again. The cracks, in all their jagged imperfection, form a map — a blueprint of lessons learned, wounds endured, and wisdom earned. I carry the scars, both literal and metaphoric, not as badges of shame but as testaments to survival. They remind me daily that pain does not have to define us; that beneath the rubble of destruction lies the potential for rebirth.

As I envision my future, it is impossible to ignore the ripple effects of my past. The ghosts of addiction, violence, and loss linger like shadows at the edge of my vision. But instead of retreating from those shadows, I now lean into them, acknowledging their weight while refusing to be dragged back under. The recovery process is not a straight line; it is a constant negotiation between the parts of me that want to surrender and those that hunger for life's promise. It is a battle fought in therapy rooms filled with raw honesty, in the quiet moments of self-reflection, and in the steady rhythm of daily rituals that anchor me in the present. Every day is a choice — a reaffirmation that I am worthy of healing, worthy of love, willing to be vulnerable, and open to transformation.

I look forward with a sense of purpose that was once foreign to me. I recognize that redemption is not a destination but a lifelong journey, one that requires patience, humility, and relentless courage. It means embracing discomfort, facing difficult truths about myself and the world, and committing to change that goes beyond personal survival to touch the lives of others. My experiences have given me a unique perspective — a clarity born from the intersection of darkness and light — and I want to use that vantage point to fight the cycles that ensnared me. I want to offer a different narrative, one that dismantles stigma and builds bridges of understanding. My story, in all its gritty realism, can serve as both a cautionary tale and a beacon for those still lost in the storm.

There is a profound responsibility that comes with surviving what so many do not. It is a charge to speak truth even when it is uncomfortable, to bear witness to the humanity in pain and struggle, and to advocate fiercely for systems that support those caught in the throes of addiction and violence. My voice emerged from the silence that once threatened to consume me, and I intend for it to be loud, unwavering, and full of empathy. I dream of reaching hearts beyond the walls of my past — connecting with policymakers who can enact real change, community leaders seeking solutions rather than scapegoats, and individuals whose lives can be transformed by knowing they are not alone.

At the same time, my vision extends intimately inward. The work of healing is as much about rebuilding relationships as it is about external activism. I am learning the art of forgiveness — forgiving myself for the wounds I inflicted in moments of desperation, forgiving those who hurt me in their own brokenness, and embracing the imperfections that make me human. Love, in all its complexity, continues to be a fragile yet powerful guide. My relationship with the Canadian escort, with all its tenderness and turmoil, was a catalyst that showed me vulnerability is not weakness but an essential element of connection. Moving forward, I seek relationships grounded in authenticity and mutual growth, where honesty is treasured over façades, and where support flows both ways.

The practical steps toward a brighter tomorrow call for sustained discipline and conscious choices. Sobriety remains my foundation, a daily commitment that requires not just willpower but a network of support, accountability, and self-compassion. Therapy continues to be a vital tool — a mirror reflecting not only past traumas but also the strengths I sometimes fail to see. Support groups remind me I am part of a larger community, a web of individuals bound by shared struggles and collective hope. Each conversation, tear, and triumph in these spaces fortifies my resolve and expands my capacity to help others.

I also envision ways to translate my journey into tangible action. Whether through speaking engagements, writing, or community programs, I want to illuminate the shadowy corners of addiction and violence that are often ignored or misunderstood. My goal is not to preach or judge but to humanize the struggles so many endure. I want to dismantle the walls of silence with stories that resonate on a fundamental level — stories that demand empathy and challenge preconceived notions. Every person I reach, every heart I touch, is a step toward a world where recovery is met with support rather than stigma, and where brokenness can be met with pathways to wholeness.

Looking even further ahead, I dream of contributing to systemic change — advocating for policies that address the root causes of addiction and violence: poverty, lack of education, inadequate mental health resources, and societal neglect. In this landscape, prevention must walk hand-in-hand with rehabilitation. The communities most affected need not only intervention programs but sustained investment in opportunity and dignity. I want to be a voice among those pushing for reform, ensuring that the lessons gleaned from my life and others like mine inform approaches to healing on a wider scale. The vision is bold, but so is the need.

Despite the challenges that pepper the road ahead, I find myself buoyed by a quiet optimism — a deep-seated belief that transformation is possible for anyone willing to confront their own darkness and reach for something better. This optimism is not naive; it is forged in fire and tempered by experience. It acknowledges the weight of history while choosing to carry a light, however small, into the uncertain future. The human spirit's capacity for resilience never ceases to amaze me, and I embrace that spirit as my guiding star.

Walking this path, I remind myself daily that healing is not linear. There will be setbacks, moments of doubt, days heavy with grief and temptation. But each obstacle is an opportunity to practice forgiveness, to deepen understanding, and to recommit to the values I now hold dear. The fractures of my past may never fully vanish, but they do not have to define the entirety of my narrative. Instead, they can coexist with stories of hope, growth, and redemption — a testimony to the complexity of the human condition.

In these reflections, I also find gratitude. Gratitude for the second chances granted, for the people who believed in me when I struggled to believe in myself, and for the unyielding spark of life that kept me moving forward. This gratitude is not passive but active—a call to give back, to serve as a guide, and to embody the change I wish to see.

My vision for tomorrow is not one of perfection but of continuous repair and building. It is a vision where my lived experience fuels empathy rather than judgment, where pain is acknowledged but not glorified, and where the broken can find pathways toward mending. I hope to be a bridge—connecting the past and the future, despair and hope, silence and voice.

As I step into this new chapter, I carry with me the knowledge that reclaiming one's life is a radical act of courage. It requires facing fears, dismantling old patterns, and daring to dream beyond survival toward purpose and joy. My journey has been a testament to that courage, and I am ready to channel it into an enduring legacy of healing and advocacy.

This brighter tomorrow calls to me every day, a beacon guiding me through the uncertainty and complexity of life. It reminds me that no matter how deep the darkness, light remains possible, and that every small act of choosing hope adds to a collective movement toward redemption. The future is still unfolding, and while I cannot predict its course, I can promise to meet it with honesty, resilience, and an open heart.

In embracing that promise, I find peace — not as a conclusion, but as a state of becoming. The journey is far from over, but for the first time, I face it not as a victim of circumstance but as an agent of my own transformation. And in that truth, I discover the most profound freedom of all.